Planting a City in the Tropical Andes

This book reveals how the nineteenth-century modernisation of Bogotá led to a transformation in the social role of plants—showing how this city located in the high altitudes of the Tropical Andes turned into a 'floristic island' formed by native, introduce, and cultivated plants.

Urbanisation is one of the main forces behind biodiversity loss. Paradoxically, the expansion of cities has endowed urban environments with a greater number of plant species compared to their surrounding areas. *Planting a City in the Tropical Andes* takes a multidisciplinary approach to shed light on the cultural and ecological mechanisms that have transformed modern cities into unique botanical spaces. By drawing upon a wide array of historical sources, this book explains how the nineteenth-century modernisation of Bogotá (Colombia) led to the replacement of traditional botanical practices with technical knowledge, which in turn endowed the city with a distinctive floristic inventory. Through a novel botanical perspective on Latin American urban history, this book uncovers how capitalist dynamics in Bogotá transformed plants into providers of clean air and water and how their use in the urban landscape contributed to the cultivation of disciplined citizenry. Placing plants at the forefront of its narrative, the book offers an original contribution to the underexplored history of horticulture in tropical Latin America. It serves as a compelling example of how the creative and conflicting forces of the Anthropocene have forged new environments and previously unseen relationships between people and plants.

This volume will be of great use to scholars and students interested in social history, urban environmental histories and cultural history.

Diego Molina is British Academy Postdoctoral Fellow at Royal Holloway, University of London and visiting reseracher at Kew Botanic Gardens. He is a botanist who turned to historical geography to better understand the changing relationships between people and plants. Before becoming a British Academy Fellow at the RHUL, he was a Landhaus Research Fellow at the Rachel Carson Center for Environment and Society, Ludwig Maximilian University of Munich.

Routledge Research on Gardens in History

Series Editors:

Karen Jones
University of Kent, UK

James Beattie
Victoria University of Wellington, New Zealand

Planting a City in the Tropical Andes
Plants and People in Bogotá, 1880 to 1920
Diego Molina

https://www.routledge.com/Routledge-Research-on-Gardens-in-History/book-series/RROGIH

Planting a City in the Tropical Andes

Plants and People in Bogotá, 1880 to 1920

Diego Molina

Routledge
Taylor & Francis Group
NEW YORK AND LONDON

First published 2025
by Routledge
605 Third Avenue, New York, NY 10158

and by Routledge
4 Park Square, Milton Park, Abingdon, Oxon, OX14 4RN

Routledge is an imprint of the Taylor & Francis Group, an informa business

ISBN: 978-1-032-07656-0 (hbk)
ISBN: 978-1-032-07658-4 (pbk)
ISBN: 978-1-003-20816-7 (ebk)

DOI: 10.4324/9781003208167

Typeset in Sabon
by Apex CoVantage, LLC

Contents

Figures

Tables

Acknowledgements

Any human creation is a natural consequence of connections, discussions and, especially, generous cooperation. Any book, then, is a collection of voices and shadows beyond the voice and shadow of its author. This book is not an exception; its existence was possible thanks to a substantial group of people and institutions.

The information that supports this book was scattered in different archives, libraries, herbariums and museums. I feel compelled to acknowledge the usually silent but pivotal role of archivists, librarians and curators in Colombia and the United Kingdom. I have special gratitude for the librarians at Kew Botanic Garden, the Biblioteca Luis Ángel Arango in Bogotá, the Museo and Biblioteca Nacional de Colombia, the Archivo General de la Nación in Bogotá and the Archivo de Bogotá.

This book is the product of my PhD in Human Geography at the University of Reading. Special gratitude needs to be extended to my two supervisors, Giuseppe Feola and Hilary Geoghegan, who had an initial influence on this book, and Luciana Martins and Matthew Gandy, my *viva* examiners who, through their constructive and challenging corrections, played a significant role in the final version of the manuscript.

Many scholars have left indelible marks on the conception and production of this book. Stefania Gallini at the Universidad Nacional de Bogotá deserves a special mention; her constant support and sharp view of environmental history in Latin America have had a decisive influence on this book. Other Colombian environmental historians and geographers also contributed with comments, constructive criticism and observations on different stages of writing. A spatial thanks to Katherine Mora and Vladimir Sánchez. Also, many thanks to the professors German Mejia Pavony and Alberto Saldarriaga Roa for their special suggestions about Bogotá´s history. I am especially grateful to professor Felix Driver at Royal Holloway, University of London; he read and corrected some parts of the manuscript, and more importantly, he has become a real mentor in my intellectual development. In addition, this book was enriched by ecological discussions with botanists and biologists such as Alvaro Idárraga, Andrés Arias, Carlos Parra-Osorio and Diego Giraldo-Cañas.

This book draws on some materials that previously appeared as published articles: 'Urban Spaces, Plants, and People in the Nineteenth-Century Bogotá (Colombia)'. *Economic Botany* 75, no. 3 (2021): 1–19; 'The Forced Retirement of a Hard Worker: The Rise and Fall of Eucalyptus in Bogotá'. *Environmental History* 27, no. 1 (2022): 58–85; 'The Genesis of Urban Flora: Colombian Cities as Islands of Disciplined Human-Plant Relationships'. *Global Environment* 16, no. 2 (2022): 325–56. I extend my acknowledgements to these journals and their editors for permission to use parts of these papers in revised form. Additionally, special thanks to the Biblioteca Publica Piloto de Medellín, the Fundación de Amigos de Bogotá, the Sociedad de Mejoras y Ornato de Bogotá, The Museo de Bogotá, the Museo del Banco de la República de Colombia and the Bourgarel Foundation for allowing me to use images of their collections.

The writing process of this book was largely enriched thanks to the Rachel Carson Center for Environment and Society and its Landhaus Fellowship. I am incredibly grateful to the fellows of this program with whom I shared six months in Glonn (Bavaria). Our debates, drinks and small adventures to Alpine lakes left a deep mark on my understanding of the human side of biodiversity.

Finally, this book was also possible thanks to the incommensurable emotional support of friends such as Fan Ya Ting, Lorenzo Todorov, Constanza Tani, Fernando Olivares and Antonio Rodríguez, who made my life wonderful when this book was just a project. Also, the transition from a manuscript to a book would be simply impossible without the multidimensional support of Zoi Krokida and the love of Fiona Kinniburgh.

The city of plants

We are never alone. No matter the size and complexity of our settlements, they are far from being exclusively inhabited by humans. Any village, town or city always involves co-existence with other creatures. It is because of our capacity to domesticate and be domesticated that, like plants, we became rooted, leaving behind our past as hunter-gatherers.[1] To guarantee a place for living, humans and their related species work together as land-crafters creating emergent physical, symbolic and semiotic ecological niches. Our modern cities are the quintessence of this interspecific co-creation of lively places. Dogs and cats as companion species, the daily use of yeast in bakeries, bacteria in sewer treatment processes, and ornamental plants decorating parks and avenues are just a few examples of organisms involved in the construction of urban environments. But cities are also inhabited by species that have acquired urbanite status without our invitation. Mammals such as rats, arthropods of different kinds, bacteria such as *Yersinia pestis*, and a myriad of hardy plants, have made of our permanent settlements their living space. This book is about those plants which have found their habitat in cities, and how the urban dwelling of plants is a historical phenomenon that reflects, in turn, our ways of conceiving, building and existing in cities.

Cities as floristic islands

For most of their life cycle, plants are not endowed with the capacity to become mobile. Hence, as rooted organism, plants cannot easily escape the development of our permanent settlements, and so with the onset of urbanisation some of them inevitably disappeared.[2] Other plants, however, are actively maintained close to us due to their medicinal, edible, magical or other properties. In the meantime, other species well-adapted to the continuous disturbances linked to the urbanisation became permanent inhabitants of towns and cities. Those are ruderal species ('weeds') and other *anecophyte* species, which, as the botanist Hildemar Scholz explains, are plants 'only found in human-created habitats.[3] Cities, then, as the result of human

DOI: 10.4324/9781003208167-1

imagination, technology and power are also a rare ecological system were humans and plants interact in ways non seen anywhere else. However, this intersection with plants in urban environments are unstable and changeable. Through our mobility we endow plants with mobility and thus we enriching our cities with species coming from distant environments. The city then act as an attraction force capable to change the biological distribution of hundreds of plant species.[4] The antagonistic interactions emerging between our domesticated plants, those native species that resist constant disturbance, and the willingly or unintentionally introduced species have made our cities 'floristic islands' that are often richer in plant species than their surroundings.[5]

Urban ecologists have attributed this phenomenon of 'floristic islands' as a consequence of various factors, including the great diversity of environments (i.e. biotopes) in the urban environment, offering suitable places for the development of plants with different ecological requirements;[6] the drastic transformation of ecosystems around cities, which has made the latter appear richer in species by contrast;[7] and the significant presence of introduced plants in cities and their antagonistic relationship with local species.[8] Such ecological accounts of the richness of plant species in cities are usually complemented with categories that imply some sense of ecological change. For example, European ecologists have coined the term *archeophytes* to describe the plants in Europe prior to the Columbian exchange; conversely, *neophytes* are all the species arriving in the same place after 1492.[9] Nevertheless, the reasons and categories used by natural scientists to explain the uniqueness of urban flora are distinctly static. Without any historical dimension, ecological explanations alone cannot account for the social, economic and aesthetic processes that have turned cities into unique botanical realms.

In order to understand how our cities became floristic islands, it is necessary to re-interpret the explanations and categories proposed by urban ecologists through the lens of historical research on the mobilities of plants, cities and people.[10] Like plants, cities are rooted; their physical infrastructures cannot easily be moved. However, this does not, of course, mean that the social and economic dynamics through which urbanism emerges cannot become mobile. Deterritorialised, the modern city travels in the guise of experts, colonial enterprise, maps, blueprints and photographs, destined to be reterritorialised in other places under very different circumstances.[11] Thus mobilised, the city creates points of contact, hybridisations and socio-ecological tensions that contribute to the diversity of green spaces and the complexity of human–plant relationships.[12] The diverse forms of urban green spaces in different modern cities reflect a complex process whereby architectural models and planning schemes are mobilised and adapted across a variety of global contexts.[13]

The expansion of particular models of the city, however, responds to the expansion of particular ways in which human bodies are/unfold in the inhabited space. This is to say that we can think of the city as an extension of the human body. When considered in that way, the city of each time has been shaped according to the organic needs and desires of each time.[14] For instance, the existence of highways or airports responds to the desire to move fast through earth. In the same vein, the presence of hospital is the built expression of our desire of being healthy. Richard Sennett explains well this expansion of the human body towards the urban space. According to this author, in the city the flesh is turned into stone.[15] This interaction also involves plants. As explained by Karen Jones, the green spaces built with plants transform urban spaces into lungs, here is the stone that turns into flesh and so does the city breathe, as its inhabitants do.[16] In this book, I try to understand how the flesh turned not into stone but morphed into gardens, parks or meadows. I assume that the diversity of green spaces and their plants, taken as an ontological extension of the human body, needs to be understood in relation to the changing ways of inhabiting urban space.

Like the bodies that inhabit it, the city grows. While growing, the city expands, carrying its inventory of urban plants and thus transforming the immediate adjacent vegetal communities. But, instead of growing like a human body, the city expands like a tree that extends its physical and symbolic branches beyond the built environment itself. The work of William Cronon and Matthew Gandy on Chicago and New York, respectively, has re-framed our understanding of the expansion of the urban.[17] Cities' interaction with their surroundings turns adjacent environments into homogenous, often environmentally degraded spaces. Starting with the historical dependence on lumber and firewood to build and cook, the growth of the modern city had profound consequences for the botanical ensemble of spaces thriving close to human settlements.[18] With the Industrial Revolution and the subsequent consolidation of urban capitalism, ecosystems around the city were progressively transformed into large monocultures. Examples of this phenomenon include the transformation of Cuban native forests into an extensive sugar cane plantation, as discussed by Reinaldo Funes, and the massive replacement of the native bush of New Zealand by a forested landscape of exotic trees following British settlement, as explored by Paul Star.[19] While such studies in environmental history are not necessarily focused on the role of urban life as the primary force in the radical transformation of peri-urban flora, reading them in this way allows us to see how, as the city expands its symbolic and material influence over a wider region, this process contributes to its development as a floristic island richer in species than its surroundings. The ecological impact of modern urbanism and its consequences on the plant communities not necessarily present in the built environment echoes the questions about the limits of the city explored by urban theorist Henri Lefebvre. Although we might be tempted to consider that the modern city is 'everywhere and in everything',[20] for the purpose of this book we need to consider how it was that the city made itself an enclave of plants, re-shaping the phenomenological and ecological experience of nature both within and beyond the tangible limits of the city.

The contrast between the botany of the built environment and its vicinity should be seen in the context of the large number of introduced species that have arrived and accumulated in modern cities over time. Gardens and other open spaces in cities are places where many medicinal, edible, ornamental or raw-material plants have been grown, part of a large-scale reorganisation of the biological distribution of thousands of individual species. Numerous academic studies highlighting human agency in the transformation of plant distributions have shed light on the expansion of crops from centres of domestication,[21] the exploitation and transportation of medicinal plants,[22] and the exploitation and transportation of raw materials such as rubber, and wood, or food.[23] Moulded by the uneven power geometry of successive waves of globalisation, a massive translocation of plants promoted the replication of particular landscapes in regions separated by thousands of kilometres. The expansion of the European flora in the Americas following its conquest by Iberians and the creation of 'Neo Europes', as originally revealed by Alfred Crosby, or the mobility of African crops as a result of the transatlantic slave trade between the sixteenth and nineteenth centuries, revealed by Judith Carney and Richard Nicholas, clearly exemplify this phenomenon.[24] However, except for a few investigations highlighting the 'tropicalisation' of English cities through the transport and maintenance of tropical plants in places such as Kew Gardens,[25] the mobility of plants linked to their use as photosynthetic raw materials in the production of modern cities, particularly in the Latin American context, has been largely ignored. Furthermore, although it is widely recognised that European ways of inhabiting and living in the city were widely used as the template in the urbanisation of the contemporary world, the role of plants in the recreation of urban spaces beyond what we recognise as Europe still needs to be investigated.[26] Doing so means explicitly addressing the question of how European models of building cities linked to particular types of plants were replicated in environments with different conditions of temperature, precipitation, sunlight and atmospheric humidity, such as those in tropical climates. Historians have underestimated how the industrial and capitalist cities on both sides of the Atlantic promoted an exchange of plants not seen since the sixteenth century. A historical approach is essential if we are to examine how recently introduced plants came to augment the botanical inventory of cities through encounters with useful and spontaneous natives, early introduced and naturalised plants, endowing each city with a particular botanical identity.

The city as a situated botanical palimpsest

Together, the mobilities of transnational urban ideas, the expansion of the city into its hinterlands and the coexistence of native and introduced plants in various urban biotopes help explain the floristic richness of urban environments. However, these mechanisms are not sufficient in themselves to provide a trajectory capable of giving shape to the historical transformation of the city into an island of plants. I use 'trajectory' here as a way of capturing the order resulting from successive economic, social and technological transformations that have modified human interaction with plants in the city. In any historical period, there are particular spaces where clearly defined relations with plants have come to the fore. Amidst continuous comings and goings, traces of these interactions remain in the landscape as evidence of the city's past in the guise of residual green spaces or as naturalised (and in some cases invasive) plant species. Understood that way, the historical transformation of the city and its plants engenders an organic and cultural palimpsest.

The idea of the landscape as a palimpsest, especially in the context of its plants, was developed in the 1940s by the cultural geographer Carl Sauer and the ethnobotanist Edgar Anderson, both from Berkeley.[27] For them, plants were living evidence of the historical evolution of places. In this context, the introduction of urban models with their associated plants and green spaces gave rise to a vegetal palimpsest. However, rather than layers of history, each of which covers the one before it, this palimpsest represents the encounters of species introduced at specific moments in the history of the city. Moreover, this botanic palimpsest is alive. The vital force of species pushes them to thrive in places beyond their biological distribution, thereby ensuring them a living space in the city. This vegetal plasticity can be witnessed, for example, through the history of the butterfly bush (*Buddleja davidii*) in the UK. Originally introduced as an ornamental plant in the 1890s from China where an earlier British explorer described the dense tangles formed by this plant as a 'famous harbourage for leopards', within a half-decade this plant had colonised every open and well-drained space in every British town.[28] This kind of biological adaptation to new environments allows many plants to persist even when the social conditions that propelled their use and introduction have disappeared. Drawing on the Dutch term *Stinzenplanten*,[29] Moritz von der Lippe explains how the existence of certain plants in Jena (Germany) can be considered an 'indicator with memory' capable of telling a detailed history about the city. Von der Lippe proposes that plants should be included in urban conservation efforts alongside monuments and heritage sites.[30] Nevertheless, although living among us as witnesses to our shared urban history, plants are usually overlooked by urban historians. We need to see the city through a botanical lens, with eyes capable of reading the flora of the city as much as its architecture. Once free from plant blindness, the remains of the botanical palimpsest can be traced not only in the landscape but also in archives, libraries, herbariums and museums, so it becomes possible to disentangle the city's trajectory as an ecologically inherited place.[31]

The superimposition of plants resulting from urban history is particular to every city. Consequently, the interactions between people, urban spaces and plants that give each city a distinct botanical identity need to be seen as spatially situated. On the one hand, the vast majority of plants around the world have a distinct spatial distribution.[32] For instance, although Manaus adopted an extensive European-like urban shape amidst the nineteenth-century rubber boom, the apophytes (understood as species occurring in undisturbed spaces which persist in anthropogenic habitats) that emerged in that city were never the same as in Paris or London.[33] Thus, as spatial assemblages of humans and other creatures, human settlements emerge from a particular botanical structure, which is in turn transformed and moulded by the introduction of urban models and plants. In short, although some investigations have shown some level of homogenisation in urban flora,[34] the initial ecological conditions historically transformed by the mobilisation of plants by people made every city a particular case without any exact equivalent. Following, I briefly outline the particular socio-ecological conditions and the botanical trajectory that make Bogotá a unique case for the study of urban nature.

Three moments of urban flora in Bogotá

The geography of urban flora is the result of interactions between the primary ecological structure and the successive historical introductions of plants and urban models. In this book, I examine the creation of modern green spaces in Bogotá during the transition from the mid-nineteenth to the twentieth century. As the botanical conditions of the nineteenth-century city were a legacy from the city's past, it is necessary to briefly present the ecological conditions of Bogotá as well as the distinct moments of botanical and urban transformation that moulded the city's urban flora.

Bogotá is a tropical Andean city with a temperature and rainfall average of 13.5°C and 900 mm, respectively. It is located on a high-altitude plateau called *Sabana de Bogotá* (Plain of Bogotá), which is mostly simply known as the *Sabana*.[35] The product of the sedimentation of an ancient lake drained in the late Pleistocene, this plain surrounds the city on its north, south and west sides. The east side of the city is flanked by the *Cerros Orientales* (the Eastern Hills), an imposing mountain chain whose orogenesis initiated 20 million years ago (mya) in the Miocene, when its summit was elevated to its current height of 2,600 metres above sea level. The complex geological history of Bogotá is the reason for the highly diverse environments surrounding the city. For example, whereas it is possible to find *páramos* (moorlands) and *bosques de niebla* (cloud forests) in the Eastern Hills, the *Sabana* is characterised by wetlands, which are locally known as *humedales*, pastures, and some dry areas.[36] This ecosystemic mosaic confers the city with high levels of biological richness and an exceptional concentration of endemic species, which, in turn, contributes to making the Tropical Andes one of the world's hotspots of biodiversity.[37]

Amidst these particular ecological conditions, the human and plant settlements in Bogotá followed three distinctive moments, which in broad terms are shared (with some local variations) with most of the cities in Latin America. The first moment was formed by the Pre-Columbian settlements and their associated plants. This was the moment when the domestication of plants such as potatoes (*Solanum tuberosum*), pineapples (*Ananas comosus*) and maize (*Zea maize*) took place in the Andes, Amazon and Central America.[38] Domesticated plants facilitated the creation of hierarchical societies, some of which developed large urban settlements where emergent interaction with plants took place in the guise of gardening and horticulture. For instance, it is widely known that highly developed horticultural practices enabled the Aztecs to endow their cities with up to five different types of green spaces comprising ludic and botanical gardens, orchards, *chinampas*, and urban forests.[39] Additionally, archaeobotanical investigations have also demonstrated that Mayan cities had a diverse array of gardens and parks whose elaborateness extended well beyond that of ordinary household horticultural spaces such as kitchen gardens.[40] This garden sophistication linked to extensive urban development was not generalised throughout the continent. In places such as today's Colombia, horticultural developments were more discrete, entuned with the smaller dimensions and complexities of its urban settlements.[41] However, this does not mean that people in pre-Columbian Bogotá did not develop particular relationships with plants. When the Spanish arrived in the city in 1537, the hierarchical society was organised in confederations formed by nucleated settlements closely linked to the cultivation of crops and the wide use of spontaneous edible plants.[42] As a fundamental moment to understand the subsequent human–plant relations in the city, this transition between the pre-Columbian Bogotá, its plants and people towards the colonial city built by Europeans is examined in depth in Chapter 1.

As a consequence of new ways of conceiving and building the city as a spatial organisation of the new colonial elite, the European conquest of Latin America inaugurated the second moment of ethnobotanical practices in Bogotá.[43] This European-like urban reality included various types of green spaces and was accompanied by the introduction of edible, ornamental and medicinal plants that were highly valued by the new social rulers and changed the floristic composition of human settlements in the

Americas.[44] Although poorly explored by scholars, the introduction and acclimatisation of plants as an integral part of European urbanism in the Americas were the urban expression of the well-studied Columbian biological exchange.[45] The result of this biological exchange of species used in cities produced an integration of European plants linked to medicine, food and gardening into American botanical knowledge.[46] During this period, pastures and monocultures of American crops such as maize and introduced ones such as wheat (*Triticum aestivum*) and barley (*Hordeum vulgare*) used to satisfy human and animal diet requirements became an essential part of Bogotá's colonial landscape.[47] The creation of the rural landscape linked to the European-like city promoted the proliferation of both native and introduced ruderal plants growing in constantly disturbed spaces. Additionally, as part of the European urbanism introduced in America, some domestic and open spaces such as patios and *solares* [kitchen gardens in backyards] became spaces for ornamental and edible plants. This combination of a city surrounded by cultures, abandoned spaces colonised by spontaneous plants and the domestic spaces made Bogotá a mosaic of green spaces. However, these spaces were not clearly defined, and the border of the city was rather diffuse.[48] The end of the Colonial period as a moment of transition is addressed in Chapter 2. There, I explain how domestic open spaces propelled particular uses of plants in a kind of botanical segregation, characterised by the possibility of access to gardens and home gardens by members of the urban elite. In Chapter 3, I examine how people at the margins of society developed particular ethnobotanical relationships with plants in non-urbanised land and nearby forests where they extracted various timber and non-timber elements that were regularly used in the city.

The urban environment of Bogotá and its plants changed little during almost three centuries of colonial control. However, beginning after the independence from Spain in 1819 and accelerating after the second half of the century, Bogotá underwent a profound transformation that would endow it with a distinctive urban physiognomy and thereby frame the third moment of human–plant interactions in the city. Capitalist dynamics entailed a new expression of urban life evident in the diversification of urban spaces. Factories, public buildings and offices became the places where life unfolded. In turn, these new spaces were interconnected by equally modern means of transport and communication such as tramways and telegraphs. This network of interconnected nodes became a dynamic scene where modern ways of life and technological innovations were displayed, attracting people from all around the country. These new urbanites were first-hand witnesses of a city undergoing a process consisting of a systematic abandonment and negation of the colonial ways of life and the embrace of new urban practices, routines and dynamics characteristic of the capitalist city.

The shape of the modern city was replicated worldwide along with the circulation of the capital, and Bogotá was no exception.[49] As was the case in many other Latin American capitals, the production of the modern city followed a given direction: the cities of the Global South adopted the model of cities created in the North, thereby reflecting that 'mobility and control over mobility both reflect and reinforce power where not everyone has an equal relationship'.[50] The mobilisation of ways of living in and building the city led to a transnational mobile urbanism that in turn engendered a cosmopolitan taste whose influence emanated from Paris. Parks, boulevards and public green spaces were fundamental aspects of Parisian urban development, and they were rapidly replicated elsewhere. In Latin America, the authorities in cities such as Buenos Aires and Rio de Janeiro hired French landscape architects such as Hubert Donat Alfred Agache, Jean Claude Nicolás Foster, and Édouard André, who introduced Parisian green spaces to Latin America.[51] Unlike these cities connected with the international trade through the Atlantic, Bogotá was a city situated in the Andes at 2,600 m.a.s.l. with a poorly developed infrastructure and less substantial participation in the international trade of raw materials. As a consequence, the urban transformation of Bogotá as a manifestation of new social, economic and political dimensions was relatively slow and on a smaller scale.[52] The nineteenth-century transformation of Bogotá to host green spaces is considered in section II of this book. Chapter 4 explores the local factors that guided the creation of green spaces in Bogotá and briefly describes the most important of them during the period analysed; Chapter 5 overviews the local and foreign gardeners that participated in the process. Chapter 6 explains the plants used in the nineteenth century, focusing on the introduction of plants by distinctive social characters. Additionally, Chapter 7 examines how all this construction of modern green spaces produced completely new and disciplined relationships between people and plants.

When gardens, promenades and green open spaces were finally established by the end of the nineteenth century, they found new ways of understanding, using and performing while endowing the city with new biotopes and ornamental plants introduced from Europe.[53] At the same time, the city expanded, and crop and pasture fields were urbanised or transformed into forestry plantations, underpinning the contrast between the built environment and its surroundings. The transformation of Bogotá's surroundings is addressed through the example of the eucalyptus plantation in Chapter 10 following an examination of the historical lack of trees in Bogotá in Chapter 8 and the creation of national symbolism around trees in Chapter 9.

This book specifically addresses the changes in human–plant interaction brought about by the transition from the colonial city to the capitalist city and how this period of change contributed to transforming Bogotá into a floristic island characterised by a botanical palimpsest in which plants and green spaces produced in different periods of its urban history were put in contact with each other. To do so, this book examines the urban ideas, urban transformations, human characters and plants that engendered this ethnobotanical and urban transition towards twentieth-century Bogotá. Focusing mainly on the century after 1819, the book brings together ideas, methodologies and practices from urban history, botany, human geography and environmental history to show how the uses and mobilities of certain species of plants can help us to understand the entangled ethnobotanical relationships characteristic of modern urban life in the Tropical Andes, one of the most biodiverse regions of the earth.

Notes

1 The domestication of animals, but especially of plants has been suggested as a key phase in the development of human settlements, see Lewis Mumford, *The City in History: Its Origins, Its Transformations, and Its Prospects*, vol. I (New York: Secker & Warburg, 1961), 10–15.
2 Using palynological evidence, Edgar Anderson explained how the appearance of the first farmers during the 'neolithic revolution' matches with an increase of pollen from cereals, 'weeds' and pioneer trees such as birch, and a reduction of oaks and other species from the forest, see Edgar Anderson, *Plants, Man and Life* (Berkeley and Los Angeles: University of California Press, 1969), 4–6.
3 Hildemar Scholz, "Einheimische Unkr ¨ Auter Ohne Naturstandorte ('Heimatlose' Oder Obligatorische Unkr ¨ Auter)," *Flora Vegetatio Mundi* 9 (1991): 105–12.
4 Anderson, *Plants, Man and Life*, 3–16.
5 P. Pyšek, "Factors Affecting the Diversity of Flora and Vegetation in Central European Settlements," *Vegetatio* 106, no. 1 (March 1993): 89–100, https://doi.org/10.1007/BF00044860; S. M. Walters, "The Next Twenty Years," in *The Flora of a Changing Britain*, ed. F. Perring (Hampton: Classey for the Botanical Society of the British Isles, 1970), 136–41; Annett Wania, Ingolf Kühn, and Stefan Klotz, "Plant Richness Patterns in Agricultural and Urban Landscapes in Central Germany—Spatial Gradients of Species Richness," *Landscape and Urban Planning* 75, no. 1–2 (February 2006): 97–110, https://doi.org/10.1016/j.landurbplan.2004.12.006.
6 Joseph Bigirimana et al., "Alien Plant Species Dominate the Vegetation in a City of Sub-Saharan Africa," *Landscape and Urban Planning* 100, no. 3 (April 15, 2011): 251–67, https://doi.org/10.1016/j.landurbplan.2010.12.012; Ian Douglas, *The Urban Environment* (London: Edward Arnold, 1983); J. Stadler et al., "Exotic Plant Species Invade Diversity Hot Spots: The Alien Flora of Northwestern Kenya," *Ecography* 23, no. 2 (2000): 169–76; W. Sukopp and M. Werner, "Urban Environment and Vegetation," in *Man's Impact on Vegetation*, ed. W. Holzner, M.J. Wereger, and I. Ikusima (The Hague: Dr. W. Junk Publishers, 1983), 247–60.
7 Kirstin Deutschewitz et al., "Native and Alien Plant Species Richness in Relation to Spatial Heterogeneity on a Regional Scale in Germany," *Global Ecology and Biogeography* 12, no. 4 (July 1, 2003): 299–311; Henning Haeupler,

"Statistische Auswertung von Punktrasterkarten Der Gef¨ Aßpflanzenflora S¨ Ud-Niedersachsens," *Scripta Geobot* 8 (1974): 1–141.

8 Myla F. J. Aronson et al., "Urbanization Promotes Non-Native Woody Species and Diverse Plant Assemblages in the New York Metropolitan Region," *Urban Ecosystems* 18, no. 1 (March 1, 2015): 31–45, https://doi.org/10.1007/s11252-014-0382-z; Pablo González-Moreno et al., "Plant Invasions Are Context-Dependent: Multiscale Effects of Climate, Human Activity and Habitat," *Diversity and Distributions* 20, no. 6 (2014): 720–31, https://doi.org/10.1111/ddi.12206; Ingo Kowarik, "On the Role of Alien Species in Urban Flora and Vegetation," in *Urban Ecology: An International Perspective on the Interaction Between Humans and Nature,* ed. J.M. Marzluff et al. (Boston, MA: Springer, 2008), 321–38; Dennis Skultety and Jeffrey W. Matthews, "Human Land Use as a Driver of Plant Community Composition in Wetlands of the Chicago Metropolitan Region," *Urban Ecosystems* 21, no. 3 (June 1, 2018): 447–58, https://doi.org/10.1007/s11252-018-0730-5.

9 Rüdiger Wittig, "The Origin and Development of the Urban Flora of Central Europe," *Urban Ecosystems* 7, no. 4 (December 1, 2004): 323–29, https://doi.org/10.1007/s11252-005-6833-9.

10 Mimi Sheller and John Urry, "The New Mobilities Paradigm," *Environment and Planning A: Economy and Space* 38, no. 2 (February 1, 2006): 207–26, https://doi.org/10.1068/a37268.

11 In their *a Thousand plateaus,* Deleuze and Guattari introduce the concept of rhizome to explain the multiples relationships between the elements of the 'plane of consistency' (the tangible reality) they explain the rhizome as an anti-genealogy of impossible dichotomic relationships. This declaration against any dualistic perspective is based on the multiplicities that can be acquired by any entity in a given time. These multiplicities are the product of a sort of movement, a sort of expansion of the component entities of the rhizome. This movement is explained through the process of deterritorialisation and reterritorialisation that I use here to explain the mobilities of cities and plants, see Gilles Deleuze and Félix Guattari, *A Thousand Plateaus: Capitalism and Schizophrenia* (London: Athlone Press, 1988). For an empirical evidence of this see, Felix Driver and David Gilbert, *Imperial Cities: Landscape, Display and Identity, Studies in Imperialism (Manchester, England)* (Manchester: Manchester University Press, 1999).

12 Ash Amin and Nigel Thrift, *Cities: Reimagining the Urban* (Cambridge, UK: Polity Press, 2002).

13 Jane Jacobs, "Ecologies of Dwelling: Maintaining High-Rise Housing in Singapore," in *The New Blackwell Companion to the City*, ed. Gary Bridge and Sophie Watson (Oxford: Wiley-Blackwell (Blackwell Companions to Geography), 2011), 79–95; John Friedmann, "Globalization and the Emerging Culture of Planning," *Progress in Planning* 64, no. 3 (October 1, 2005): 183–234, https://doi.org/10.1016/j.progress.2005.05.001; Anthony King, *The Bungalow: The Production of a Global Culture* (London: Routledge & Kegan Paul, 1984); Peter Kraftl, "Geographies of Architecture: The Multiple Lives of Buildings," *Geography Compass* 4, no. 5 (2010): 402–15, https://doi.org/10.1111/j.1749-8198.2010.00332.x.

14 José Luis Pardo, *Las Formas de La Exterioridad* (Valencia, España: Pretextos, 1992).

15 Richard Sennett, *Flesh and Stone: The Body and the City in Western Civilization* (London: Faber and Faber, 1994).

16 Karen R. Jones, "'The Lungs of the City': Green Space, Public Health and Bodily Metaphor in the Landscape of Urban Park History," *Environment and History* 24, no. 1 (February 1, 2018): 39–58, https://doi.org/10.3197/096734018X15137949591837.

17 William Cronon, *Nature's Metropolis: Chicago and the Great West, 3. Print* (New York: Norton, 1992); Matthew Gandy, *Concrete and Clay: Reworking Nature in New York City/Matthew Gandy, Urban and Industrial Environments* (Cambridge, MA: MIT Press, 2002).

18 William Cronon, *Nature's Metropolis: Chicago and the Great West*, 1st ed. (New York: W. W. Norton & Company, 1992); John Perlin, *A Forest Journey: The Story of Wood and Civilization* (New York and London: W. W. Norton & Company, 1989); Joachim Radkau, *Wood: A History* (Cambridge, UK: Polity, 2012); Charles Watkins, *Trees, Woods and Forests: A Social and Cultural History*, 1st ed. (London: Reaktion Books, 2014).

19 Reinaldo Funes, *From Rainforest to Cane Field in Cuba: An Environmental History since 1492, Envisioning Cuba* (Chapel Hill: University of North Carolina Press, 2008); Paul Star, "Tree Planting in Canterbury, New Zealand, 1850–1910," in *Trees, Themes in Environmental History 5*, ed. Sarah Johnson (Cambridge, UK: The White Horse Press, 2015), 101–19.

20 'To what extent is each city expanded?' and 'Where are the limits of the city?' were questions often posed by Henry Lefebvre. Amin and Thrift (2002, 1) respond to these questions to some extent when they state that *city is everywhere and in everything*. Does the contemporary city leave spaces free of its influx? For example, the intensive and destructive gold extraction in the Amazon Basin is not the extension of a single city; however, it can be conceived of as a consequence of the 21st century's overwhelming global expansion of urban life. In this sense, this book aligns with the statement proposed by Amin & Thrift (2002, 1).

21 See Dorian Fuller and Lucas Leilani, "Adapting Crops, Landscapes, and Food Chioses: Patterns in the Dispersal of Domesticated Plants across Eurasia," in *Human Dispersal and Species Movements. From Prehistory to Present*, ed. Nicole Boivin, Rémy Crassard, and Michael Petraglia (Cambridge: Cambridge University Press, 2017), 304–31; Carl Ortwin Sauer, *Agricultural Origins and Dispersals, Bowman Memorial Lectures* (New York: The American Geographical Society, 1952); Nikolai Vavilov, "The Origin, Variation, Immunity and Breeding of Cultivated Plants," trans. Starr Chester, *Chronica Botanica* 13 (1951): 1–366.

22 See Mauricio Nieto, *Remedios Para El Imperio. Historia Natural y La Apropiación Del Nuevo Mundo*, 2nd ed. (Bogotá: Universidad de los Andes, 2006); Rohan Roy, *Malarial Subjects: Empire, Medicine and Nonhumans in British India, 1820–1909, Science in History* (Cambridge: Cambridge University Press, 2017).

23 See Warren Dean, *Brazil and the Struggle for Rubber: A Study in Environmental History, Studies in Environment and History* (Cambridge: University Press, 1987); Stephen L. Harp, *A World History of Rubber: Empire, Industry, and the Everyday*, 1st ed. (Chichester, West Sussex: Wiley Blackwell, 2015). Perlin, *A Forest Journey*; Radkau, *Wood.* John Soluri, *Banana Cultures: Agriculture, Consumption, and Environmental Change in Honduras and the United States* (Austin: University of Texas Press, 2006); Funes, *From Rainforest to Cane Field in Cuba.*

24 Judith Ann Carney and Richard Nicholas Rosomoff, *In the Shadow of Slavery: Africa's Botanical Legacy in the Atlantic World* (Berkeley: University of

California Press, 2009); Alfred W. Crosby, *Ecological Imperialism: The Biological Expansion of Europe, 900–1900, Canto* (Cambridge: Cambridge University Press, 1993). See also Nathan Nunn and Nancy Qian, "The Columbian Exchange: A History of Disease, Food, and Ideas," *The Journal of Economic Perspectives* 24, no. 2 (2010): 163–88; Manuel Patiño, *Plantas Cultivadas y Animales Domésticos En América Equinoccial*, vol. III, 8 vols. (Cali: Imprenta Departamental, 1967); Charles Mann, *1493: How Europe's Discovery of the Americas Revolutionized Trade, Ecology and Life on Earth* (London: Granta Books, 2012).

25 Andrew Hassam, "Portable Iron Structures and Uncertain Colonial Spaces at the Sydenham Crystal Palace," in *Imperial Cities*, ed. Felix Driver and David Gilbert (Manchester: Manchester University Press, 1999), 174–93; Rebecca Preston, "'The Scenery of the Torrid Zone': Imagined Travels and the Culture of Exotics in the Nineteenth-Century British Gardens," in *Imperial Cities*, ed. Felix Driver and David Gilbert (Manchester: Manchester University Press, 1999), 194–211.

26 The lack of research on the mobilities of plants underlying the creation of green spaces in Latin American cities contrasts with a more solid research on the use of introduced plants in the production of urban spaces in Asia, see for instance James Beattie, "Colonial Geographies of Settlement: Vegetation, Towns, Disease and Well-Being in Aotearoa/New Zealand, 1830s-1930s," *Environment and History* 14, no. 4 (2008): 583–610; James Beattie and Joanna Boileau, "'Cultivated with Great Carefulness': Chinese Market Gardening, Urban Food Supplies and Public Health in Australasia, 1860s–1950s," *New Zealand Journal of History* 54, no. 2 (2020): 57–85; Robert Peckham, "Hygienic Nature: Afforestation and the Greening of Colonial Hong Kong," *Modern Asian Studies* 49, no. 4 (2015): 1177–209.

27 Carl Sauer, "Foreword to Historical Geography," *Annals of the Association of American Geographers* 31, no. 1 (1941): 1–24, https://doi.org/10.2307/2560961; Anderson, *Plants, Man and Life*.

28 D. F. Owen and W. R. Whiteway, "*Buddleia davidii* in Britain: History and Development of an Associated Fauna," *Biological Conservation* 17, no. 2 (February 1, 1980): 149–55, https://doi.org/10.1016/0006-3207(80)90045-2.

29 This term refers to a group of plants that originally appeared in a region only as an imported ornamental plant species in estates, farmyards, presbytery gardens and the like, and maintained themselves as feral, see Piet Bakker and Evert Boeve, *Stinzenplanten* (Zutphen: Uitgeverij terra, 1985).

30 Moritz von der Lippe, "Vegetation as Testimony: Botanical Traces of Urban Past," in *The Botanical City*, ed. Matthew Gandy and Sandra Jasper (Berlin: Jovis, 2020), pp. 46–53.

31 Kevin Laland, Blake Matthews, and Marcus W. Feldman, "An Introduction to Niche Construction Theory," *Evolutionary Ecology* 30, no. 2 (April 1, 2016): 191–202, https://doi.org/10.1007/s10682-016-9821-z.

32 James Brown and Mark Lomolino, *Biogeography*, 2nd ed. (Sunderland, MA: Sinauer Associates, 1998); Alexander von Humboldt, *Tableaux de la nature* (Librairie des sciences naturelles et des arts illustrés, 1865), http://archive.org/details/tableauxdelanat04humbgoog.

33 M. Rikli, "Die Anthropochoren und der Formenkreis des Nasturtium palustre DC," *Ber. Schweizer Bot. Ge* 8 (1903): 71–82, https://doi.org/10.5169/SEALS-13224.

34 Ingolf Kühn and Stefan Klotz, "Urbanization and Homogenization—Comparing the Floras of Urban and Rural Areas in Germany," *Biological Conservation*,

Urbanization 127, no. 3 (January 1, 2006): 292–300, https://doi.org/10.1016/j.biocon.2005.06.033.

35 Ernesto Guhl, *Colombia: bosquejo de su geografía tropical*, vol. 2, 2 vols. (Bogotá: Instituto Colombiano de Cultura, Subdirección de Comunicaciones Culturales, 1975).

36 Byron Calvachi, "La Diversidad Bogotana," *Revista La Tadeo*, no. 67 (2002): 89–98.

37 Norman Myers et al., "Biodiversity Hotspots for Conservation Priorities," *Nature; London* 403, no. 6772 (2000): 853–58, https://doi.org/10.1038/35002501.

38 Dolores R. Piperno, "The Origins of Plant Cultivation and Domestication in the New World Tropics: Patterns, Process, and New Developments," *Current Anthropology* 52, no. S4 (2011): S453–70, https://doi.org/10.1086/659998; Vavilov, "Review of The Origin, Variation, Immunity and Breeding of Cultivated Plants."

39 Susan Toby Evans, "Aztec Royal Pleasure Parks: Conspicuous Consumption and Elite Status Rivalry," *Studies in the History of Gardens & Designed Landscapes* 20, no. 3 (September 1, 2000): 206–28, https://doi.org/10.1080/14601176.2000.10435621; José Morales, "Jardines Prehispánicos de México En Las Crónicas de Indias," *Archivo Español de Arte* 77, no. 308 (2004): 351–73.

40 Javier Caballero, "Maya Homegardens: Past, Present and Future," *Etnoecológica* 1, no. 1 (1992): 35–54; Barbara Stark, "Ancient Open Space, Gardens, and Parks: A Comparative Discussion of Mesoamerican Urbanism," in *Making Ancient Cities. Space in Early Urban Societies*, ed. Andrew Creekmore III and Kevin Fisher (Cambridge, UK: Cambridge University Press, 2014), 370–406.

41 Carl Henrik Langebaek, *Regional Archaeology in the Muisca Territory: A Study of the Fúquene and Susa Valleys, University of Pittsburgh Memoirs in Latin American Archaeology*, no. 9 (Pittsburgh: Santfé de Bogotá: University of Pittsburgh; Universidad de los Andes, Departamento de Antropolgía, 1995).

42 See for instance Hector Flores et al., "Andean Root and Tuber Crops: Underground Rainbows," *HortScience* 38, no. 2 (2003): 161–67, https://doi.org/10.21273/HORTSCI.38.2.161; see also Marienne Schrimpff, "En busca de los primeros agricultores del altiplano cundiboyacense," *Manguaré*, no. 5 (1987): 99–125.

43 José Luis Romero, *Latinoamérica Las Ciudades y Las Ideas, Historia y Cultura* (Buenos Aires: Siglo veintiuno, 2014).

44 Patiño, *Plantas Cultivadas y Animales Domésticos En América Equinoccial.*

45 Crosby, *Ecological Imperialism.*

46 Francisco Aceituno and Juan Martín, "Plantas Ameriandias En La Mesa de Los Primeros Europeos En Panamá Viejo," *Latin American Antiquity* 28, no. 1 (March 2017): 127–43, https://doi.org/10.1017/laq.2016.9.

47 Katherine Mora, "Adaptación de Sociedades Agrarias a La Variabilidad Climática. Sabana de Bogotá, Andes Orientales Colombianos, 1690–1870" (Doctorado en Historia, Universidad Nacional de Colombia, 2017).

48 Diego Molina, "Urban Spaces, Plants, and People in the Nineteenth-Century Bogotá, Colombia," *Economic Botany* 75, no. 3–4 (December 2021): 268–86, https://doi.org/10.1007/s12231-021-09524-5.

49 David Harvey, *The Urbanization of Capital, Studies in the History and Theory of Capitalist Urbanization*, vol. 2 (Oxford: Blackwell, 1985).

50 Arturo Almandoz, "Modernización urbanística en América Latina. Luminarias extranjeras y cambios disciplinares, 1900–1960," *Iberoamérica* 7, no. 27 (2007): 59–78; Arturo Almandoz Marte, *Planning Latin America's Capital*

Cities, 1850–1950, Planning, History, and the Environment Series (London: Routledge, 2002); David Morley, *Home Territories: Media, Mobility and Identity, Comedia* (London: Routledge, 2000).

51 Hofer Andreas, *Karl Brunner y El Urbanismo Europeo En América Latina* (Bogotá: El Áncora Editores, 2003); Sonia Berjman, *Plazas y parques de Buenos Aires: la obra de los paisajistas franceses, André, Courtois, Thays, Bouvard, Forestier, 1860–1930*, 1st ed. (Buenos Aires: Fondo de Cultura Económica de Argentina, 1998); Sonia Berjman and Anatole Tchikine, "Landscape Architecture in Latin America: Nineteenth and Twentieth Centuries," *Studies in the History of Gardens & Designed Landscapes* 39, no. 3 (July 3, 2019): 175–77, https://doi.org/10.1080/14601176.2018.1561817.

52 Germán Mejía, *Los Años Del Cambio. Historia Urbana de Bogotá, 1820–1910, Colección Biblioteca Personal*, Segunda edición (Bogotá: CEJA Pontificia Universidad Javeriana Instituto Colombiano de Antropología e Historia, 2000).

53 Claudia Cendales Paredes et al., *La Vida Privada de Los Parques y Jardines Públicos: Bogotá 1886–1938* (Bogotá: Instituto Distrital de Patrimonio Cultural, 2020); Hugo Delgadillo, *El Parque Del Centenario En Bogotá: Transformación Urbana, Itinerario y Significado* (Bogotá: Universidad Nacional de Colombia, 2017).

Section I

Early encounters between people and plants

1 The genesis of Bogotá's mixed flora

At the beginning of the sixteenth century, the European conquerors were unable to envision the dimensions of the land they were occupying in South America. The invaders had defeated the Incas in 1533, and they were looking for an easy link between the land of Perú and the Caribbean. In April 1536, the Spanish Conqueror Gonzalo Jiménez de Quesada departed from the Caribbean city of Santa Marta with 800 men looking for this connection. After eight months, the expedition had not found any trace of the Incas' gold. However, they found scattered inhabitants who, communicating as they could, informed the foreigners about an extensive network of paths that the locals used in the trade of salt. 'Trade' meant 'wealth' in the minds of the newcomers, so they headed uphill following the route.[1]

On March 1537, after struggling for more than a year, only 173 members of the original troop finally arrived at a populated land on a high plateau situated atop the Andes at 2,600 meters above sea level. This was, Bacatá, the pre-Columbian Bogotá. After having experienced unimaginable mishaps, deaths and disease due to the harsh weather, hunger, parasites and extreme fatigue, the chronicler Juan de Castellanos celebrated the Europeans' arrival to a place of serenity, describing it as a 'Land of gold and well supplied/Land to build an eternal house/Land of abundant food/Land of big civilisations/Plainland of dressed people. . . . Holy land, calm and serene/ Land that puts an end to our sorrow'.[2]

The 'dressed people' mentioned by Castellanos were part of a well-organised society whose gold, emerald, and finely cotton-woven blankets made the Spanish forget the still distant Perú. Instead, they passed the following months exploring the area, making war with the indigenous inhabitants and looting the land.[3] When the locals' resistance was finally suffocated, the Spanish secured territorial control. Now it was possible to find a city. Following an elaborated Foundational Act, on the 6th of August of 1538, Gonzalo Jiménez de Quesada dismounted from his horse, and pulling up some grass and strolling around the area, said that 'he was taking possession of that place and land in the name of the most undefeated

DOI: 10.4324/9781003208167-3

Emperor Charles the Fifth'.[4] In doing so, the conquerors gave birth to the city of Santafé de Bogotá.

However, that land did not belong to the distant kingdom of Spain. It was the homeland of a great civilisation celebrated by Castellanos: the Muiscas Confederation. The Muiscas' ancestors were hunter-gatherers who had inhabited the area as early as 12,000 years ago (ya) and acquired some early agricultural practices around 5,000 ya and perfected them by 3,000 ya, when palynological records document systematic agriculture practices in the highland plateau.[5] Agricultural expansion facilitated an increase in the human population, which was accompanied by more complex social organisation. The Muisca nation must have been numerous when the Spanish arrived by the end of the sixteenth century—estimates indicate that between 600,000 and one million inhabitants were living in an area of 600 square acres.[6] In 1562, when contact with European diseases had begun to decimate the native population, the chronicler Pedro Aguado observed that Tunja (another town in the plateau 150 km from Bogotá) was heavily populated by residents who gathered in 'endless numbers' every four days in the occasion of the market.[7]

This large population of indigenous people was politically organised into *cacicazgos*, governance units gathered through matrilineal blood whose influence expanded across various settlements of different dimensions. Muiscas' towns comprised a group of *bohios* (circular thatched huts) built close to each other to form triangles or strength lines. Between the houses were areas of transit, communal buries and home gardens.[8] Maize, which was the base of the Muiscas' diet, was profusely grown in the home gardens as well as in nearby cultivation areas. However, this grain was not the only crop that they cultivated. The first Europeans who settled in Muiscas' lands observed that people in Bogotá consumed a tuber they called *turmas*, which tasted like radishes.[9] As in other Andean cultures, tubers played a central role in the diet: *ibias* (*Oxalix tuberosa*) *ullucos* (*Ullucus tuberosus*), *cubios* (*Tropaeolum tuberosum*) and potatoes were widely used by people in Bogotá. The first two were possibly used by the first inhabitants of the plateau as early as 5,000 ya.[10] In addition, Muiscas cultivated beans (*Phaseolus vulgaris*) and various types of *cucurbitas* (gourds), and they had access to a wide spectrum of Andean fruits growing in perturbed areas such as *uchuva* (*Physalis peruviana*), *curuba* (*Passiflora tripartita*) and *cerezo* (*Prunus serotina*).[11] Besides the cultivated or freely growing plants around their settlements, the Muiscas' society had developed a sophisticated trade network that enabled them to complement their diet with plants from warmer areas, such as avocado (*Persea americana*) and pepper (*Capsicum annuum*).[12]

The European's arrival drastically transformed the Muiscas' social organisation and their ways of living with plants. The new rulers brought their god, their social structures, their values and prejudices, their cities and their plants. Because growing their population would guarantee the Spanish political, economic and spiritual control of the territory, the creation of stable settlements was the most critical part of the conquest campaign. Without cities, the long-term exploitation of nature and people would be impossible. In other words, urban settlements worked as a link between the conquest and the colonial land.[13] That is why in 1552, the chronicler Francisco López de Gomara, who wrote a history of *Las Indias* without being there, observed that 'Those [conquerors] that do not populate will not conquer well. [They] will not conquer the land and will not convert people [to Christianity]. The main aim of any conqueror is to populate'.[14] However, while the foundation of cities was a relatively easy process, their maintenance over time was not always so. Harassed by hostile indigenous inhabitants and lacking sufficient knowledge to use the local resources, many cities were abandoned or relocated several times.[15] Hence, the prevalence of the European settlements was only possible due to the local existence of a subordinated indigenous population and their knowledge on how to interact with plants, animals and other resources. The first European settlement in Bogotá clearly exemplifies this relation of dependence. The conquerors' city was initially merely an encampment formed by 12 huts and one rustic thatched church built following the local architecture of the Muiscas' *bohios*. The latter was a product of the Muiscas' culturally accumulated knowledge on how to turn plants into building materials.

The early city of Bogotá was aspirational. To materialise the European settlement as the theatre of power and the space where new social relationships unfolded was necessarily an enormous effort of the settlers [*vecinos* in Spanish].[16] According to the *Leyes de Indias*, they had to settle in the recently conquered territory to obtain the benefits of the land; moreover, the settlements would only receive Crown recognition if they were properly inhabited by Spanish people.[17] The existence of Bogotá as an urban European-like settlement was not made possible through the mere physical presence of Europeans on the Muiscas' land. Rather, the city's long-term existence was mainly guaranteed through the modification of local nature, and primarily through the introduction of plants and animals that allowed the newcomers to reproduce their customs and ways of life.

Domestic plants and animals common to the conquerors were a cornerstone of the initial European-like establishment.[18] Quesada's expedition included horses and dogs as animals of war and transport. As was usual in any conquest campaign, they also transported pigs, goats and lambs. Known as *carnada*, these animals were used as livestock during the conquest campaign and as an initial supply of European domestic animals in Muiscas' territory.[19] It is not clear whether some of this *carnada* was still alive when Quesada's expedition arrived in the Muiscas' land. Notably, chronicles observe that other conquest expeditions were undertaken soon after Quesada's introduced the first domesticated species from Europe to the area. For example, Sebastian de Belalcázar's campaign, which departed from Quito with 150 men (all of which survived the journey), introduced the first pigs in Bogotá. On the other hand, the less fortunate campaign headed by the Swabian Nicolas de Federmann is known as the introducer of the first hens to Muiscas' territories from Venezuela, despite the fact that when the 160 men of the expedition arrived in Santafé de Bogotá, they were ill and covered in the furs of the wild animals they had hunted to tackle famine during their journey.[20]

There is no clear information about seeds introduced in Bogotá by the city's European founders. However, considering the work and time needed to grow crops such as wheat and barley, it seems certain that the first conquerors initially relied on indigenous crops such as maize. In any case, it is well documented that during their conquest campaigns and first settlements in America, Europeans devastated the local fields by devouring what they found; thus, their tastes became early accustomed to the indigenous foods.[21] However, Europeans in the recently founded Santafé de Bogotá could not only rely on the abundance of unfamiliar tropical foods. In January of 1540, a settlement entourage headed by Jerónimo Lebron arrived in Santafé, bringing with them some elements that defined the European character of the settlement. Clothes, wine, and some handicrafts were part of this group. They also brought the seeds of undetermined cereals (possibly wheat and barley) and introduced more domestic animals, including mares.[22]

Along with and of equal or greater importance to the seeds and animals, the settlement group headed by Lebron came with six women to aid the task of populating the Muiscas' territory with Spanish blood. With the exception of the handcrafters who embodied particular skills needed to replicate European ways of life (e.g. the construction of churches), many of the new settlers did not arrive to work. A significant number of the first European inhabitants of the city proved to have aristocratic blood, which implied a natural distaste for any manual labour. Additionally, other settlers had earned rights over land and people through their active participation in the conquest campaign.[23] Thus, many Spanish arrived in Bogotá to take over portions of the looted land and live at the expense of the indigenous population. The availability of resources and the possibility of growing European plants were crucial in the development of the city; however, as proved in the cases of Cuzco and Mexico City, the future of this *ciudad hidalga* (aristocratic city), as named by José Luis Romero, was only guaranteed by a massive indigenous working force able to plant, care, cultivate and transport the fruits of the recently introduced species as well as the now profitable native plants.[24]

Europeans sought to develop a trading system that elevated the domestic economy above simple subsistence. The prospect of profiting from commerce motivated many of them. To accomplish this, there was a strenuous effort to keep indigenous people herded together. The defeated Muiscas were not permitted to live in Santafé de Bogotá and were instead segregated in nearby *Pueblos de Indios* (Indian towns). Following the grid-like European layout, these towns were conceived by Spanish authorities as efficient ways of controlling the recently subdued people. On the one hand, religious leaders insisted that gathering indigenous people together made them easier to indoctrinate. Additionally, civil authorities viewed these settlements as a way to accelerate acculturation and the introduction of civil and control practices.[25] Nonetheless, one of the main motivations for putting the indigenous population together was rooted in economic interests. The Pueblos de Indios were used as economic nodes where the Spanish controlled production through the so-called *La Encomienda* (the assignment), a rigid and brutal economic and social control scheme of semi-slavery and vassalage relationships in which the Spanish Crown assigned groups of indigenous people whose souls needed saving by outstanding Europeans. Such a master of persons was called *El encomendero*, and in exchange for his spiritual services for the indigenous people, he received the products of their labour in the guise of tribute. Conquerors' descendants and other high-ranking members of society used this human and environmental exploitation system built upon the indigenous tribute of both traditional and new raw materials as a way of creating wealth.[26] Keeping indigenous people together under the *Encomienda* guaranteed strong political control while also ensuring that a human workforce was available to exploit agricultural products, forest biodiversity and mineral richness.

Although these settlements were physically separated, botanical practices started to amalgamate inside both the Pueblos de Indios and the European city within a few decades. Forced to produce introduced species, the indigenous had to develop a particular relationship with plants thus far ignored by them. Wheat, barley, and occasionally chickpeas and fava beans were included in the tribute that often also contained traditional indigenous cultivars such as maize and beans.[27] As part of their engagement with new plants and agricultural practices, it is highly probable that the indigenous also had to learn how to deal with unwillingly European-introduced 'weeds' such as broadleaf plantain (*Plantago major*) and dandelion (*Taraxacum officinale*).[28] Moreover, although the historical record is poor, it is also highly probable that indigenous farm labourers used crop surpluses to enrich their traditional diets based on maize, calabashes, beans, *ullucos*, *ibias* and potatoes.

The most direct relationship with plants was embodied in the indigenous people, who had to plant, nurture and harvest both introduced and native plant species. Additionally, they maintained ethnobotanical practices linked to magical, medicinal, and aromatic and poisonous plants.[29] This close relationship between indigenous people and plants found a public expression in Pueblos de Indios markets. A decade after the founding of Bogotá, the Spanish monarch ordered that 'In every Pueblo de Indios it was compulsory to have a market and squares endowed with *mantenimientos* (groceries) for the Spanish and indigenous travellers could buy with their money what they could need'.[30] On the other extreme of the society, the inhabitants of the Spanish city did not participate in the cultivation and retailed commercialisation of edible and other useful plants. The Spanish descendants' interactions with plants remained in the realms of alimentation and commerce as products of indigenous tribute. However, this does not mean that indigenous crops were not part of their daily lives. For instance, although the inhabitants of the Spanish city showed a preference for a Mediterranean diet based on cured meats and imported goods such as olives, constraints on commerce forced them to consume local edible plants.[31] Additionally, indigenous servants slowly introduced ingredients such as *achiote* (*Bixa orellana*) which was extensively used as a food dye, and *guasca*, used as a condiment, thus transforming the elites' diets and tastes.[32]

The undifferentiated use of plants in the Pueblos de Indios and the Spanish city anticipated the extensive biological and cultural miscegenation that took place in later decades. As a consequence of the indigenous demographic collapse by the end of the seventeenth century, the *encomienda* system began to decay, and the settlements that stemmed from it underwent a significant transformation. The difference between the Spanish city and Pueblo de Indios became diffuse with the increase of the mestizo population.[33] In the *ciudad mestiza* (Creole city), native and introduced plants and hybrid botanical practices created a landscape where urban and rural environments interacted without defined borders. In total, 85 years after Bogotá's founding, Fray Pedro Simón described how the land two leagues (approximately 10 km) around Bogotá was occupied by many 'mill farmhouses with wheat, barley, and corn'. Simón explained that it was also possible to find 'Native and Castilian legumes', evidencing how Bogotá had become a place of botanical encounter. However, when detailing the crops and vegetables cultivated in Santafé, Simón only mentions the presence of cabbage (*Brassica oleracea*), lettuce (*Lactuca sativa*), chicory (*Cichorium intybus*), fava beans, carrots (*Daucus carota*), spinach and chard (*Beta vulgaris*), all plants introduced by Europeans. Interestingly, Simón's testimony sheds light on the naturalisation process of certain introduced species, such as mustard (*Brassica rapa*), which according to him had been found in Bogotá '*su centro y propia tierra*' (its centre and its own land). Simón's chronicle also evidences how the built environments of the colonial city hosted other species growing in what we could consider urban environments. Simón shows the existence in Bogotá of 'many and very good orchards within the town, with the abundance of vegetables' that, besides containing the entire spectrum of vegetables cultivated around the city, also had herbs such as camomile and spearmint, which 'works as *juncia* (herb) to cover the soil in any occasion'. Finally, Simón also explained how these orchards/home gardens were rich in fruits such as figs (*Ficus carica*), quinces (*Cydonia oblonga*), grapes (*Vitis vinifera*), apples (*Malus domestica/Malus pumila*), plums (*Prunus domestica*), almonds [Undetermined species], pears of the *cermeña* variety (Varieties of *Pyrus communis*), and peaches (*Prunus persica*). However, once again, the chronicler only mentions those fruits introduced from Europe, ignoring the vast diversity of fruits native to the Andes that certainly grew in the city.

The interactions with plants in Bogotá during most of the colonial period are still waiting for more detailed studies. However, we know that by the eighteenth century, progressive cultural amalgamation made it impossible for most people to distinguish between the *legumbres de la tierra y de Castilla* mentioned by Fray Pedro.[34] By that time, the Pueblos de Indios had been absorbed by the European city, and some plants, such as the mustard, had become fully naturalised. The convergence of the two cities gathered indigenous and European plants and ethnobotanical practices in the same space; this does not mean, however, that the eighteenth-century city's ensemble of useful plants was static. As demonstrated in the case of Mexico City, people linked to ecclesiastical power introduced various species of ornamental and edible plants.[35] Although the role of monks and nuns as active introducers of plants in Andean cities is poorly understood, they must have had a key role, as many of the cultivated spaces in the colonial city were orchards, gardens and home gardens within monasteries and convents. Equally, as I will show in the last section, in the framework of the Bourbonic reforms, local authorities established promenades (*alamedas*) that slightly modified the organisation of plants within the urban spaces of Bogotá by the end of the eighteenth century.

The relations with plants in Bogotá underwent a new transformation once full independence from Spain was achieved in 1819. The redefinition of the social order undermined the church's authority in favour of republican institutions. Soon after the independence, gardens, orchards and kitchen gardens in religious buildings such as convents and monasteries disappeared while these buildings were used to host the new republican institutions. At the same time, the integration into the international capitalist trade as raw material suppliers promoted the creation of a new bourgeois class that replaced the power niche left after independence.[36] This modernisation was an eminently urban phenomenon.[37] The city became the epicentre of modernity and the place where elites' vision of 'progress' was crystallised. Encompassing substantial physical (i.e. infrastructure projects) and symbolic (i.e. discourses) changes, these transformations engendered an entirely new urban experience and a new way of interacting with plants.[38] The following sections present the changing relationships with plants in Bogotá after Colombian independence. Specifically, I highlight how the second half of the nineteenth century saw the emergence of the urban plant in concomitance with vast urban transformations that occurred when the city hastily passed from the preindustrial period to a world dominated by machines and the hegemony of market ideology.[39]

Notes

1 Germán Mejía, *La Ciudad de Los Conquistadores 1536–1604, Serie Historia de Bogotá* (Bogotá: Pontificia Universidad Javeriana, 2012).
2 Juan De Castellanos, *Elegías de varones ilustres de Indias por Juan de Castellanos, Biblioteca Autores Españoles*, vol. IV, 2nd ed. (Madrid: Imprenta y Estererotipía de M. de Rivadeneyra, 1852), 310.
3 Mejía, *La Ciudad de Los Conquistadores 1536–1604.*
4 Pedro Simón, *Noticias Historiales: De Las Conquistas de Tierra Firme En Las Indias Occidentales, Biblioteca Banco Popular*, vol. III (Bogotá: Banco Popular, [1675] 1981), 301.
5 Gonzalo Correal and Thomas van der Hammen, *Investigaciones arqueológicas en los abrigos rocosos del Tequendama: 12.000 años de historia del hombre y su medio ambiente en la altiplanicie de Bogotá, Biblioteca Banco Popular* (Bogotá: Banco Popular, 1977); Thomas Van der Hammen, "La Estratigrafía e Historia Del Neogeno y Cuaternario de La Cuenca Alta Del Río Bogotá," *Análisis Geográficos (IGAC)*, no. 26 (2003): 101–20.
6 J. Michael Francis, "Población, enfermedad y cambio demográfico, 1537–1636. La demografía histórica de Tunja: una mirada crítica," *Fronteras de la Historia* 7 (December 10, 2002): 13–76, https://doi.org/10.22380/20274688.682; Ernst Röthlisberger, *El Dorado: Estampas de Viaje y Cultura de La Colombia Suramericana, Edición facsimilar de la primera edición Colombiana, vol. I, Apuntes Maestros* (Bogotá: Universidad Nacional de Colombia. Rectoría, 2016), 170.
7 Carl Henrik Langebaek, *Mercados, Poblamiento e Integración Étnica Entre Los Muiscas Siglo XVI, Colección Bibliográfica* (Bogotá: Banco de la República, 1987), 47–102.
8 Francisco Romano and Lorena Lemus, *Nueva Esperanza. 2000 Años de Historia Prehispánica de Una Comunidad En El Altiplano Cundiboyacense* (Tunja: Universidad Pedagógica y Tecnológica de Colombia—UPTC, 2018), 60.
9 Gonzalo Fernández de Oviedo, *Historia general y natural de las Indias, Biblioteca de Autores Españoles no. 117 a 121* (Madrid: Ediciones Atlas, 1959), 110–12.
10 Marienne Schrimpff, "En busca de los primeros agricultores del altiplano cundiboyacense," *Manguaré*, no. 5 (1987): 99–125.
11 Edible species common in perturbed areas can be considered as synanthropic species adapted to live close to human settlements. However, we still know little about these spaces as part of the Andean settling process. For a European example, see Rüdiger Wittig, "The Origin and Development of the Urban Flora

of Central Europe," *Urban Ecosystems* 7, no. 4 (December 1, 2004): 323–29, https://doi.org/10.1007/s11252-005-6833-9.

12 Langebaek, *Mercados, Poblamiento e Integración Étnica Entre Los Muiscas Siglo XVI*, 47–102.

13 German Colmenares, *Historia Económica y Social de Colombia 1537–1719*, 2da ed. (Medellín: Editorial La carreta & La Oveja Negra, 1975), 18–19; Mejía, *La Ciudad de Los Conquistadores 1536–1604*.

14 The original text in Spanish is as follows: *Quien no poblare, no hará buena conquista, y no conquistando la tierra, no se convertirá la gente; así que la máxima del conquistador ha de ser poblar,* ver Francisco López de Gómara, *Historia general de las Indias y Vida de Hernán Cortés* (Caracas: Fundacion Biblioteca Ayacuch, 1979), 67.

15 Romero, *Latinoamérica Las Ciudades y Las Ideas*, 63.

16 Mejía, *La Ciudad de Los Conquistadores 1536–1604*.

17 Colmenares, *Historia Económica y Social de Colombia 1537–1719*, 19.

18 The big picture of the Ecological conquest has been largely debated and well explained since the work of Alfred Crosby see Alfred W. Crosby, *Ecological Imperialism: The Biological Expansion of Europe, 900–1900, Canto* (Cambridge: Cambridge University Press, 1993).

19 José Avellaneda, "La vida cotidiana en la Conquista," in *Vida pública en las ciudades republicanas*, ed. Beatriz Castro (Bogotá: Grupo Editorial Norma, 1996), 16–56; José Ignacio Navas, *La expedición de Gonzálo Jiménez de Quesada al mar del sur y la creación del Nuevo Reino de Granada* (Bogotá: Banco de la República, 1995), 10.

20 José Vergara and José Gaitán, *Almanaque de Bogotá, Guia de Forasteros Para 1867* (Bogotá: Imprenta de Gaitán, 1866), 50.

21 Francisco Aceituno and Juan Martín, "Plantas Emeriandias En La Mesa de Los Primeros Europeos En Panamá Viejo," *Latin American Antiquity* 28, no. 1 (March 2017): 127–43, https://doi.org/10.1017/laq.2016.9; Avellaneda, "La vida cotidiana en la Conquista."

22 Pedro Ibáñez, *Las Crónicas de Bogotá y de Sus Inmediaciones*, vol. I (Bogotá: Imprenta de la Luz, 1891), accessed July 23, 2016, www.banrepcultural.org/blaavirtual/historia/crBogotá/17.htm.

23 Romero, *Latinoamérica Las Ciudades y Las Ideas*.

24 A good example of the Columbian exchange in the creation of the Spanish word in Andes is offered by Daniel Gade in his study about the introduction of plants and animals in the sixteenth century Perú, see Daniel W. Gade, "Particularizing the Columbian Exchange: Old World Biota to Peru," *Journal of Historical Geography* 48 (April 1, 2015): 26–35, https://doi.org/10.1016/j.jhg.2015.01.001; Romero, *Latinoamérica Las Ciudades y Las Ideas*.

25 Colmenares, *Historia Económica y Social de Colombia 1537–1719*, 52; Carl Henrik Langebaek, *Conquistadores e indios: la historia no contada, Historia*, Primera edición (Bogotá, DC and Colombia: Debate, 2023), 120–23.

26 Colmenares, *Historia Económica y Social de Colombia 1537–1719*, 52.

27 Ibid., 146–48.

28 Anderson, *Plants, Man and Life*, 3.

29 An example of the traditional use of plants by indigenous in the Colonial city is presented by Marta Herrera through the history of an indigenous woman who killed another one using poisoned herbs, see Marta Herrera Angel, "Te echo estas yerbas para que te muras," *Anuario Colombiano de Historia Social y de la Cultura*, no. 21 (January 1, 1993): 157–69.

30 Juan Friede, *Documentos inéditos para la historia de Colombia (1509–1550)*, vol. X (Madrid: Editorial Arco, 1955), 155.
31 Anthony McFarlane, "El Comercio En La Vida Económica y Social Neogranadina," in *Historia de La Vida Cotidiana En Colombia*, ed. Beatriz Castro (Bogotá: Grupo Editorial Norma, 1996), 363–90.
32 Aída Martínez, "La Vida Material En Los Espacios Domésticos," in *Historia de La Vida Cotidiana En Colombia* (Bogotá: Grupo Editorial Norma, 1996), 337–62.
33 Margarita Garrido, "La vida cotidiana y pública en las ciudades coloniales," in *Historia de la vida cotidiana en Colombia* (Bogotá: Grupo Editorial Norma, 1996), 131–58.
34 Martínez, "La Vida Material En Los Espacios Domésticos."
35 Gonzalo Tlacxani, "The Convents of Mexico City in the Colonial Biological Exchange: Notes for Colonial Environmental History from the Sixteenth to the Eighteenth Centuries," *Global Journal of Human-Social Science: B Geography, Geo-Sciences, Environmental Science & Disaster Management* 18, no. 2 (2018): 1–13; Tarsicio Salcedo, "Huertas virreinales la multiplicidad de su función," *Esencia y espacio* (2012): 14–33.
36 Eduardo Posada Carbó, ed., *Colombia. La Apertura al Mundo, América Latina En La Historia Contemporánea*, vol. III, V vols. (Barcelona: Fundación MAPFRE & Taurus, 2015).
37 Luciana L. Martins and Mauricio A. Abreu, "Paradoxes of Modernity: Imperial Rio de Janeiro, 1808–1821," *Geoforum, Urban Brazil* 32, no. 4 (November 1, 2001): 533–50, https://doi.org/10.1016/S0016-7185(01)00012-4; Mejía, *Los Años Del Cambio*.
38 The idea of modernisation as a group of projects, discourses and experiences is based on John Jervis, *Exploring the Modern: Patterns of Western Culture and Civilization* (Oxford: Blackwell Publishers, 1998); Sarah Radcliffe, "Geographies of Modernity in Latin America: Uneven and Contested Development," in *When Was Latin America Modern? Studies of the Americas*, ed. Nicola Miller and Stephen M. Hart, 1st ed. (New York: Palgrave Macmillan, 2007), 21–48.
39 Martínez, "La Vida Material En Los Espacios Domésticos."

2 Urban spaces, cultivated plants and people

The post-independence Bogotá had become a gathering place for cultivated and introduced plants existing alongside wild and naturalised species. Echoing this botanical integration, people in Bogotá, now labelled as Colombians, were living together in the republican city. However, living together did not mean a complete and undifferentiated integration. Although plants were capable of dispersing and growing whenever was possible, as a matter of fact, plants, people and their cultural interactions remained separated in clearly defined hierarchical urban spaces inherited from the colonial social organisation. Through the examples of ornamental, edible and wild plants, this chapter will shed light on how interactions with plants during most of the nineteenth-century Bogotá were defined mainly by the structure of the city, which was, in turn, a reflection of the inherited social structure.

The patios as domestic gardens in Bogotá

During the period that covered the colonial and early republic, gardens were scarce in Bogotá, and, when they existed, were mainly limited to private spaces in patios. A possible explanation for this lack of urban spaces decorated with plants is the lack of ostentation places as a manifestation of the local powers. Thus, any crown officials such as viceroys or judges assigned to the colonial Santafé de Bogotá knew they would be in a remote and poor province completely lacking the lively environment enjoyed in places such as Potosi, Cuzco, Lima, or Mexico City. The recreation of royal manners and the aristocratic style of life simply was not possible in this city. Throughout colonial history, Santafé occupied a secondary position against Mexico City and Lima, the two viceroyalties from which the Spanish crown had to administer a territory of roughly 10 million km Square.[1] The colonial city of Santafé was subordinated to the Lima viceroyalty until 1717, when it was named the capital of the new viceroyalty of Nueva Granada. Its inferior legal category made this city a poorer settlement where the ostentation places for the ruling powers were almost inexistent.[2] As a

DOI: 10.4324/9781003208167-4

consequence, few gardens of pleasure or ostentation existed in the Colonial city.

The few gardens of the city were located in patios of the convents and monasteries clusters as well as in the patios of private houses occupied by high members of the Crown administration. These colonial high-rank civil servants did not have palaces or palace-like constructions that reflected the social status of this bureaucratic body. They did not have gardens such as the Chapultepec in Mexico City, and the only symbol of social status was summarised in occupying a large dimension house with rammed earth walls and a tiled roof. In Santafé, this house, paradoxically known as *Palacio de los Virreyes* (Viceregal Palace), was initially located in the southeast corner of the Plaza Mayor (Central Square) and had several patios for their private use and the establishment of some gardens. Unfortunately, the Viceregal Palace was seriously affected by an earthquake in October 1743 and then completely destroyed by fire in May's 1786. After the fire, the last colonial viceroys had to use as their residence a rented property belonging to the prominent Saenz de Santamaria family.[3] As it was common for the members of the elite, this house had also a series of patios (see Figure 2.1), of which it is possible at least one was turned into a domestic garden. As part of the private realm of the civil and religious elite, the inwalled gardens in patios did not promote the development of high-skill gardening and horticulture practices. Contrarily, for most of the colonial history, these gardens were built and cared through vernacular knowledge inherited from Spain, a knowledge that extended beyond the independence in 1819 and existed as late as the beginning of the twentieth century.

Because the patio was the urban space where many gardens existed in the colonial Santafé and in the republican Bogotá, it is essential to contextualise its existence in the city. The *Casas de patios* (patio houses) was a dominant cultural trait among many of the ancient inhabitants of the Iberian Peninsula.[4] The Islamic occupation of the Iberian Peninsula between the eighth century and the end of the fifteeth century further enriched the architectural tradition of patios, adding new symbols related to introversion, sensuality and the sacred meaning of culture.[5] Consequently, the patios acquire particular importance within the Castellan and Sevillian. For them, the patio was related to the concept of a spiritual connection through spatial openness toward the sky.[6] Inquired in the early twentieth century about the importance of the patio, a Sevillian responded: 'The patio isolates you from the world and only admits one neighbour: the heaven, the supreme God's house. . . . And in this patio isolation among flowers—hydrangeas [*Hydrangea macrophylla*], geraniums [*Pelargonium* spp.], foxgloves [*Digitalis purpurea*]—the Sevillian is driven by a weird happiness'.[7]

Figure 2.1 Patios in a colonial house used by the Nueva Granada's last viceroys in 1810, 2022, digital photograph.

Source: Author's image

The traditional idyllic use of patios as a place to connect with heaven surrounded by flowers had to face an adaptive process under the social–environmental reality of Bogotá. The Sevillian house of patios, when translated into America, and especially to the cold weather of the highland Andes, was transformed in shape and function.[8] For example, in Andalucía, the family used both floors of the house, living on the ground floor during the summer and on the first floor during the winter (due to the thermal isolation); but, because of the high humidity of the ground floor in Bogotá, the family generally used to live in the first floor, leaving the ground floor for the servants or other tenants.[9] Additionally, the patios' size was reduced and multiplied to satisfy the house owner's economic needs. Some of the patios, for instance, were devoted to storing the product of the encomienda or used as stables.[10] Notably, the houses of patios were constructions of considerable since that demanded a significant economic effort. Consequently, these houses were inhabited by the most prominent members of the society, and their gardens (when they existed) were a privilege of a reduced number of citizens (see Figure 2.2). In turn, while the number and size of patios were

Figure 2.2 Anonymous, *La Quinta Segovia* [general view of a colonial patio and its garden], 1904, photograph.

Source: Courtesy of the Sociedad de Mejoras y Ornato de Bogotá

all elements of social distinction among Bogotá's elite, the rest of the people in the city had to live in small, bad airs renting rooms or in thatched huts like their Muiscas ancestors.[11]

The patio's adaptation in size, shape and use is not equal to asserting that gardens in patios were absent in Bogotá. As a matter of fact, considering the urban nature of the conquest, it is highly possible that most of the early introduced ornamental plants have found their 'natural place' in pots within patios. The lack of historical evidence prevents us from understanding the disruptions and/or continuities in the use of the patio as a garden in the transition towards the republican city from 1819 onwards. However, Soledad Acosta de Samper (1833–1913), who is considered the most important nineteenth-century Colombian woman writer, left a piece of extremely rare evidence concerning the domestic use of plants in Bogotá's patios during the post-independence period. In her book *Mi Madrina* (My Godmother), written in 1870, Acosta described the Godmother patio/garden of her childhood (ca. 1840) thusly:

> I still remember that place as it was before. I can see the tall rosemary always in blossom, the Quito tomato [*Solanum betaceum*], the cherry tree [*Prunus domestica*] and the gorse [*Ulex europaeus*] surrounded by hollyhocks [*Alcea rosea*] growing in between the loose rocks, alongside many Jericho roses [*Rosa gallica*] in whose shadow grew a fluffy carpet made out of chamomile [*Matricaria chamomilla / Chamaemelum nobile*] and

> tinged *trinitarias* [*Bougainvillea glabra*]—replaced now by non-fragrant violets [*Viola tricolor*]—and strawberries [*Fragaria vesca*]. whose leaves I always stopped to admire whenever I encountered the fruits. There were geraniums of open mound and pigeon feet classes [*Pelargonium* spp.]. In the stone bench that separated the patio and the corridor there were bowls whose flowers were more carefully tended: they had blue and white *farolitos* [*Campanula medium*], yellow *ridículos* [*Calceolaria* sp.], dark and fragrant *pompas* [undetermined species], *boton de oro* [*Achyrocline satureioides*], *pajaritos* of all colours [*Epidendrum cf. ibaguense*]. The columns were climbed by *donzones* [*Lathyrus odoratus*] and *madreselvas* [*Lonicera japonica*]. And finally, on the floor by four moulded clay pots—they were used to collect water during winter—, you could find flowerpots and broken plates in which some seedlings grew waiting to be transplanted at the appropriate time. Almost all the flowers loved by my godmother back in the days have lost their importance, and only can be found in the old-fashioned gardens of those decadent Santafereños.[12]

The use of plants in the ornamentation of patios can be thought of as the prolongation of a Hispanic inheritance. The Europeans not only brought their religion, private property, and patriarchal society to America but also carried their sensibilities and particular aesthetic approaches toward the 'natural' world in general and to plants in particular. As a consequence, ornamental plants were introduced quite early during the conquest campaign. For instance, lilies and roses were introduced to the Americas in 1520 with the migrations of Spanish women or entire families.[13] It seems that the remarkable isolation of Bogotá across colonial times, linked to the poorly developed spaces of public ostentation, did not radically change the composition of the flora used for the ornamentation of private spaces, which indicates a massive overlooking of the potential presented by some native species of plants. In 1899, José María Cordovez underscored this botanical bias when describing the types of flowers used as ornamentation for the *bailes* (home parties) in colonial Bogotá:

> Regarding flowering plants, it is necessary to confess it, the number known by them (ancient Bogotanians) was reduced. By that time, there was no notion about the immense richness and variety of the Colombian flora. The Castilla roses [*Rosa gallica*], used nowadays only to produce *colirios*, single and small carnations [*Dianthus caryophyllus/Dianthus barbatus*], poppies [*Papaver* spp.], nasturtiums [*Tropaeolum majus or Consolida ajacis*], *pajaritos*, satin flowers, hollyhocks, Guadalupe parasites [local orchids], Madonna lilies [*Lilium candidum*], and some other species, were the main material from which the flower decorations were made by those days.[14]

The predilection for European ornamental plants as raw materials in the construction of domestic spaces within the patios was still a constant during the second half of the nineteenth century. The French Édouard François André, recognised botanist, plant hunter, and horticulturist, described one of the patios-gardens he found during his stay in Bogotá between 1875 and 1876:

> A detailed survey of one garden allows me to understand one of Bogotá's traits. The plan expresses better than any description their *gardening for entertainment* in cold weather. It shows a parallelogram in the house's patio divided into rectangles and diagonal lines. It also has paths paved with baked clay tiles and a central drain to gather the rain-water. The vegetal 'furniture' is almost always the same. The rose bushes, rarely trimmed, blossom all the year through. Sometimes (in this garden) they plant a Colombian walnut [*Junglans neotropica*] in the centre, this tree has beautiful leaves and thick fruits covered by a grey pericarp. The list of flowering plants randomly placed cover violets [*Viola odorata*], fuchsias [*Fucshia* spp.], geraniums, asparagus [*Asparagus officinalis*]—As ornaments!—verbenas [*Aloysia citriodora*], balms [*Melissa officinalis*], lilies [*Lilium* sp.], wallflowers [*Erysimum* sp.], poppies, knight's spurs [*Consolida* spp.], lupines [*Lablab purpureus* or *Lupinus* sp.], spindles [*Euonymus japonicus*], common marigold [*Calendula officinalis*], arum-lily [*Zantedeschia aethiopica*] and primrose (*Primula* sp) always flowered. *The efforts in choosing flowers brought from Europe are evident everywhere*. [Emphasis added].[15]

André's views of the Bogotanians' effort to use plants from Europe do not represent a change in the directionality of the plant's introduction started in the sixteenth century. Nevertheless, this systematic selection of plants from abroad was not a response to a transatlantic migration of people who rebuilt a nostalgic representation of their homeland overseas as the first Europeans in America did. Instead, this willing maintenance of European flora is a reflection of the mimicking enactment, which would be a constant characteristic in the creation of Bogotá's nature. However, despite any Europeanisation attempt of gardens through the consistent use of introduced plants, local botanical diversity was impossible to exclude, and, consequently, these patio-gardens hosted a certain number of native species present in the city since the early sixteenth century. An example of this is the Quito tomato in the godmother's patio described by Soledad Acosta de Samper, the Guadalupe parasites mentioned by José María Cordovez, and the *Junglands neotropica* observed by André. All of them are native plant species of the Tropical Andes broadly used by indigenous and present in the colonial, republican and modern Bogotá. In short, plants in the private spaces of a minority living in houses with patios reflect the city's hybrid flora.

In contrast to the diversity of plants found in the patio-garden, the people responsible for planting and caring for these plants were considerably reduced. Using the Acosta descriptions of a patio, it may be evident that women, but specifically, the housemistresses such was the Soledad Acosta

Godmother, were responsible for the creation, ornamentation, and maintenance of the patio-garden. Women embodied domestic horticultural practices such as letting seedlings grow in suitable broken pots as they await a good moment to be transplanted. The relationship between enclosed patios and women is not surprising if we consider the patio's existence by itself has been partially linked with the need to maintain a vigorous defence of the private world against the interference that the outside world could present in the woman's house. According to Amos Rapoport,

> the fact that the court house is very much the same in Greece, North of Africa, and Latin America, suggests that the latter relates to some social factor, which may be the extreme need for privacy for women who are cloistered. The windows and roofs of these court houses are designed to prevent anyone from intruding into the intimacy of the house.[16]

Rapoport's statements might be gender-essentialist, as it ignores, for example, weather considerations, local variations, and the fundamental issue of women's agency; nonetheless, it is widely accepted that the role of women in colonial Colombian society, at least among the wealthy and middle classes, was eminently linked to the domestic sphere and men were related to the public.[17] As a consequence of the spatial segregation of genders, women devoted most of their time engaged in domestic activities, among which, as Acosta de Samper's description suggests, gardening practices played an important role. Additionally, women had a key role in the social recognition of a family since the public recognition rested not only on the physical structure of the house (e.g. size and the number of patios) but also on the organisation, neatness and decoration.[18] The access to the most private sector of the house (rooms and central patio) was reserved by people close to the family but also was a place where people from the same family class had access when, for example, other women visited the housemistress to chat or sew. Thus, the patio could be used by the housemistress as a space of social differentiation between people of the same social class.

Furthermore, as exemplified in the case of the early twentieth-century Medellín, the patio-garden was used as a symbol of social distinction that differentiated people living in houses of patios from other members of the society. In this city, in the frame of the Fiesta de las Flores (Flower's Celebration) celebrated then in May, people from the elite open their doors and windows, thus allowing others to see from outside the 'exquisite' ornamentation of their patios endowed with *bifloras* (*Rhododendron indicum*), *violetas de los Alpes* (*Cyclamen persicum*), and various palms planted in clay pots.[19] Thus, the organisation, management, and transformation of patios into a particular sort of garden space could have represented a unique chance for the mistress of the house to express her aesthetic sensibility, as well as a way of socialising with other gardening enthusiasts through the exchange of seeds and seedlings, as still happens today among many women in rural environments close to Bogotá (observ. pers.).

The reduced mobility of women in the early nineteenth-century Bogotá paradoxically produced a particular kind of mobile garden characterised by easily transported plants planted in pots that could be interspersed with the more fixed planting spaces. The house's mistress was in charge of the use, care, and control of a large number of objects, ranging from grills, bellows, mortars, and brooms to other, more distinctive objects like dressers, settees, and jewellery boxes.[20] Given her characteristic as the administrator of non-human elements inside the house, mistresses such as Acosta de Samper's godmother were able to link inanimate objects such as broken plates, pots, and vessels with plants. In doing so, she was able to produce a very distinctive type of private garden endowed with certain levels of internal mobilities that allowed her, for example, to move plants between different areas of the house and change their localisation according to the changes in sunlight throughout the year.[21]

The accelerated urbanisation process that Bogotá underwent during the end of the nineteenth century resulted in the rapid disappearance of green spaces such as patios and solares that had promoted particular uses of plants in the city.[22] In an attempt to host a growing population, many houses with patios were subdivided, and their resulting rooms, known as *tiendas* [small, dark, and poorly ventilated rooms], were rented to needy families.[23] This change in the domestic sphere around patios led to the abandonment of the gardens previously planted on patio surfaces. Soledad Acosta lamented the loss of private green spaces when she wrote that '[a]lmost all the flowers loved by my godmother back in the days have lost their importance'. Such statements highlight the disappearance of a very particular type of private urban garden that scholars have not yet fully explored. Additionally, although the historical evidence concerning the floristic composition of these gardens is severely limited, their disappearance probably entailed the loss of uncommon plants used in the decoration of these places, which still occurs in contemporary patios in Central American cities.[24]

Surviving patios and some new ones built in the so-called Republican-style houses had an adaption to the modern times' manners. Closing the nineteenth century, the sky leaking through the patio was not necessarily the representation of domestic heaven anymore. However, the sky had something to do with the common usage given to patios. Being the most brilliant part of the house was the only suitable place to gather the family around a new device: the photographic machine. Being part of a wedding, birthday celebration or individual portraits, plants in pots or planted in the ground could be appreciated in the background of the images, making patios a place of confluence between photosynthetic and photographic processes (see Figure 2.3). This photographic *infraknowledge* hidden in each image

Figure 2.3 Fotografía Rodríguez, *Isaac Restrepo* [and his family posing in the patio], 1911, photograph.

Source: Courtesy of the Biblioteca Pública Piloto de Medellín

suggests that the patio-garden occurred in Bogotá as much as this type of house existed in the city.[25] Furthermore, the graphical evidence suggests how the domestic flora in inner patios remained in the city as late as the twentieth century, thus coexisting with public parks and public gardens and making the city a conglomerate of biotopes richer in plants than the surroundings.

El solar: a condensed rural land in the city

The differentiation between the urban and rural realms in the nineteenth-century Bogotá was not clearly sketched. One of the most significant characteristics of this lack of physical differentiation was the large presence of agricultural practices within the city. An example of this rural–urban interwoven reality was expressed in the use of growing plants in an unbuilt plot adjacent to houses called *solares* [kitchen gardens in backyards].

The solares were the seminal unit of occupation through which the city grew after being founded by Jiménez de Quesada. Following the settlement regulations established in the *Leyes de Indias*, after founding a city, the conqueror had to distribute square plots of land named solares among the people who had participated in the conquest.[26] The assignation of solares in the city represented the religious and political powers that would rule the new settlement.[27] The centre of this new urban universe was the main square where the ruling institutions of the new society such as the church, the *cabildo* [the administrative building] and the *Casa del conquistador* [Conquerors' house] were located. Around this core, the city was organised in a concentric hierarchy, such that the distance to the main square was proportional to the social positions of those entitled to have a solar. High-ranking soldiers, members of the new bureaucratic body (e.g. judges and notaries) and recognised members of the society, such as those holding any kind of nobiliary title, received a solar. The Leyes de Indias stressed that they were required to build their houses in their assigned plots within a year or else lose their right to the land.[28] As a result, people first built temporary houses using straw and planks, mimicking the Muiscas' vernacular buildings. Once they had established possession over the given solar, the conquerors built their permanent dwellings. Establishing domain over the solar through its edification went beyond its physical use as shelter. In the medieval-born Spanish tradition implanted in America, any individual should be part of a recognisable house and solar (*casa y solar conocido*); this meant that any man or woman, whether they be noble or servant, must belong to a given place. Thus, while promoting a sense of belonging to the community, this attachment of people to land also entailed an efficient way of political control.[29]

By the nineteenth century, the solares in the city had lost their original meaning. Often flanked with wide walls of rammed earth, the solar was a kind of backyard, or more accurately, it was a sort of condensed rural environment used to supply domestic requirements in terms of food, storage, and animal keeping. Although Bogotá was surrounded by arable land, the solar persisted as the most immediate productive and multifunctional piece of land. In contrast to patios, which were places devoted to contemplation through the aesthetic organisation of ornamental plants, the solares, which were dedicated to the cultivation of edible and medicinal plants, were places linked with human bodily functions, specifically those related to the digestive processes.

As an extension of the colonial dynamic, the nineteenth-century house in Bogotá was a scenario of the coexistence of two adjacent but contrasting worlds. If the main rooms and patios were the territory of the house's mistress and family, the solares and related areas like the kitchens were the domain of servants and non-human domestic actors such as dogs, cats, gens, chickens, horses, and, of course, varying kinds of plants.[30] Despite the scarcity of available information concerning the social dynamics within the solares, they were one of the most vivid places where the strict norms that ruled the central part of the house were blurred.

One could consider the solares as being the most permeable part of the house, a porous membrane constantly crossed by flows of matter and energy embodied in groceries, gossip, deliveries, and varying kinds of human and non-human social actors that—unwelcomed to enter via the main door—accessed the house from the backyard. For instance, it was through the solar entrance that the chocolate grinder woman came from time to time to produce the chocolate balls to be drunk by the family during dinner, as did the laundry woman who delivered cleaned clothes. The visits made by servants' lovers occurred there, and it was the playground used by the youngest members of the household. Not all of the solares were surrounded by big rammed earth walls; some merely were merely bordered by a wooden fence or a line of giant planted *cabuya* plants (*Agave* spp. and *Furcrea* spp.). As a consequence, mischievous children frequently found success in entering the solares looking for fruits or vegetables to steal. Likewise, wandering cows or horses, very common street inhabitants, often went into the solares in search of food.[31]

As mentioned earlier, whereas patios were places for relaxation and contemplation, solares were associated with most organic human body functions and needs, which was reflected in the kinds of plants that grew in these spaces. Solares were versatile spaces used simultaneously as medicinal herb gardens, kitchen gardens, and orchards (see Figure 2.4). Considering that medicinal herbs were 'abundant in any vegetable garden',[32] and judging from the 'astonishing' number of medicinal plants still present in Bogotá's markets and their therapeutic uses,[33] it is easy to speculate that many well-known plants like chamomile and dandelion, which were undoubtedly thriving in solares, were used as part of a vernacular pharmacopoeia in the initial home treatments of less serious ailments such as diarrhoea, fever, stomach aches, or the flu.

The second group of plants growing in solares was edible plants used to feed the culinary needs of the generally adjacent kitchen of firewood ovens. In 1852 during his pass in Bogotá, the Scottish botanist Isaac Holton observed that in the house of a Don *Fulano* (Mr Nobody) there was a 'second unpaved patio' with 'a fig-tree (*Ficus carica*), a *papaya* (*Vasconcellea pubescens*), plumes and a minute apple-tree half dead with cold'.[34] Likewise, Felipe Pérez mentioned how in the city it was also possible to find some other cultivated fruit trees such as *duraznos* (*Prunus persica*) and *cerezas* (*Prunus serotina*), growing along with other spontaneous fruits plants such as *uva camarona* (*Macleania rupestris/Gaultheria sclerophylla*), *uva de anís* (*Cavendishia bracteata*) *mortiños* (*Hesperomeles goudotiana / Miconia* sp./Ericaceae spp.), *moras* (*Rubus* spp.), *esmeralda* (*Miconia albicans*) and *uchuvas* (*Physalis peruviana*).[35]

Figure 2.4 Anonymous, *Outlook of Las Aguas Neighbourhood* [houses with cultivated solares], 1890, photograph.

Source: Courtesy of the Archivo Fotográfico de José Joaquín Herrera Pérez, Fundación de Amigos de Bogotá

According to Isaac Holton, some other vegetables and fruits cultivated in the solares included potatoes and esculents such as *arracachas* (*Arracacia xanthorrhiza*), *cubios*, (*Tropaeolum tuberosum*), and *ocas* (*Oxalis tuberosa*), spinach (*Spinacia oleracea*), *curuba* (*Passiflora tripartita*), anise (*Pimpinella anisum*), *arverja* (*Pisum sativum*), fava beans (*Vicia faba*) and other beans (*Phaseolus* sp.), *maize*, *pepinos* (*Cucumis sativus*), and pumpkins (*Cucurbita maxima*).[36] Finally, Soledad Acosta observed the presence of *chisguas* (*Canna indica*), from whose corms have been traditionally ground to obtain a flour used in the production of bakery products.[37]

Compared with other Colombian cities such as Medellín or Cali, whose warm weather enables an incredible diversity of fruit trees to thrive, Bogotá had relatively fewer types of such trees. However, European-introduced apple, pear, fig, and cherry trees were commonly present in the orchards of the solares, along with native *uva camarona* (*Macleania rupestris*), *mortiños* (Colombian blueberries, *Vaccinium meridionale*) or *papayuela* (mountain papaya, *Vasconcellea pubescens*) trees.[38] Fruits were both consumed fresh and used in the preparation of desserts, as is still very typical in Bogotá's culinary tradition.[39] Thus, the presence of plants of different origins reflects the cultural hybridism expressed in culinary form. By the eighteenth century, European and American tastes, ingredients and culinary practices had essentially melded, making it difficult for most people to distinguish between what had been introduced and what was native.[40]

Whereas the house's mistress produced diverse spaces with plants of various origins, women servants in the house of the elite did the same in solares. Women's hands played a prominent role in the integration of ethnobotanical practices and plants in the solares. Servant women were responsible for keeping the fire burning and transforming fruits and vegetables from the market and the solar into dishes, and women moulded the endless reproduction of hybrid Latin American culture through culinary practices. Hence, this mixed culinary knowledge was driven by the development of a hybrid botanical knowledge about the fruits, vegetables, grains, and staples used in the kitchen, many of which were cultivated and maintained by women in urban solares.[41]

As occurred with the patios, the urbanisation of the late nineteenth-century Bogotá implied the disappearance of the solares and their edible and medicinal plants. Indeed, the economic limitations that prevented the city's expansion engendered a concentration of people in the traditional quarters, and open spaces such as solares were rapidly urbanised.[42] The disappearance of the solares expelled the last traces of the rural life that had coexisted with urban practices throughout Bogotá's history. Nevertheless, although the urbanisation of solares may have eradicated the cultivation of edible plants such as the *cubios*, potatoes, or Andean papaya described by Isaac Holton, it did not undermine the use of other plants such as medicinal herbs. Studies of present-day Bogotá have illuminated the 'astonishing' uses of herbs in the city.[43] Other recent works have reported that despite the lack of large solares, lower-income people have continued cultivating medicinal and aromatic herbs in pots and small domestic gardens.[44]

Notes

1 Romero, *Latinoamérica Las Ciudades y Las Ideas*, 46–68.
2 Martínez, "La Vida Material En Los Espacios Domésticos," 337–62.
3 Eduardo Posada, *Narraciones*, 2nd ed. (Bogotá: Villegas Editores, 1988), 141–60.
4 The Phoenician society, whose palaces had patios, had arrived in the region by 700 BCE, and when the Romans later won control of the territory, they imposed their own way of life, including patio houses inherited from Hellenic culture. The Islamic occupation of the Iberian Peninsula between the 8th century and the end of the 5th century further enriched the architectural tradition of patios, adding new symbols related to introversion, sensuality, and the sacred meaning of culture, see Juan Chávez, *Casa, Hogar y Cielo. Las Tres Miradas Debreyanas Sobre El Espacio Doméstico En El Valle de Los Aburráes, Colección Arquitectura, Hábitat y Urbanismo* (Medellín: Universidad Nacional de Colombia (Sede Medellín). Facultad de Arquitectura, 2017), 89.

5 For a rich explanation of the importance of intramural gardens in the Muslim world and its influences in the south of Spain see Penelope Hobhouse, *Plants in Garden History: An Illustrated History of Plants and Their Influence on Garden Styles, from Ancient Egypt to the Present Day* (London: Pavilion Books, 1997), 41–69.
6 Antón Capitel, *La Arquitectura Del Patio* (Barcelona: Gustavo Gili, 2005); Chávez, *Casa, Hogar y Cielo*; Marta Beatriz Silva, "La vivienda a patios de origen hispánico y su difusión en Iberoamérica," in *Actas III* (Sevilla: Congreso Internacional del barroco Americano: Territorio, Arte, Espacio y Sociedad. Universidad Pablo de Olavide, 2001), 70, https://dialnet.unirioja.es/servlet/articulo?codigo=4088197.
7 Joaquín Romero, *Los Jardines de Sevilla (Curso de Conferencias Sobre Urbanismo y Estética En Sevilla)* (Sevilla: Academía de Bellas Artes de Santa Isabel de Hungría, 1955).
8 Amin and Thrift, *Cities: Reimagining the Urban*; Eugene McCann and Kevin Ward, *Mobile Urbanism: Cities and Policymaking in the Global Age, Globalization and Community*, vol. 17 (Minneapolis: University of Minnesota Press, 2011); Silva, "La vivienda a patios de origen hispánico y su difusión en Iberoamérica."
9 James Steuart, talking about the house where he was hosted in Bogotá in 1836 explains how 'the best dwelling-houses are those of two stories high, where the inmates make use of the lower parts only for stables or storehouses.', see James Steuart, *Bogotá in 1836–7: Being a Narrative of an Expedition to the Capital of the New-Granada and a Residence There of Eleven Months* (New York: Harper & Brothers, 1838), 110.
10 Silva, "La vivienda a patios de origen hispánico y su difusión en Iberoamérica."
11 About the adaptation of the patios to the American reality, see Margarita Garrido, "La vida cotidiana y pública en las ciudades coloniales," in *Historia de la vida cotidiana en Colombia* (Bogotá: Grupo Editorial Norma, 1996), 135.
12 Soledad Acosta de Samper, "Mi Madrina," in *Recuerdos de Santafé, Libro al Viento Capital* (Bogotá: Intituto Distrital de las Artes-Idartes-, [1870] 2013), 17–32.
13 Víctor Patiño, *Aspectos Históricos Sobre Los Recursos Naturales y Las Plantas Utiles En Colombia* (Bogotá: Instituto Colombiano de Cultura, 1977), 357.
14 José Cordovez, *Reminiscencias Escogidas de Santafé y Bogotá* [Recurso Electrónico] (Bogotá: Ministerio de Cultura: Biblioteca Nacional de Colombia, 1899), 19.
15 Édouard André, "L'América Équinoxiale (Colombie-Équateur-Pérou)," *Le Tour Du Monde. Nouveau Journal Des Voyages* 35 (1878): 161–177.
16 Amos Rapoport, *House Form and Culture* (Englewood Cliffs, NJ and Hemel Hempstead: Prentice-Hall, 1969), 65.
17 Martínez, "La Vida Material En Los Espacios Domésticos," 43–84.
18 María Astrid Ríos Durán, "La casa santafereña, 1800–1830: en el camino hacia la intimidad, el confort y la domesticidad," *Anuario Colombiano de Historia Social y de la Cultura*, no. 34 (January 1, 2007): 43–84.
19 Diego Molina, *Los Árboles Se Toman La Ciudad. El Proceso de Modernización y La Transformación Del Paisaje En Medellín, 1890–1950* (Medellín: Editorial Universidad de Antioquia, 2015).
20 Martínez, "La Vida Material En Los Espacios Domésticos"; Josefa Acevedo, *Tratado sobre economia domestica para el uso de las madres de familia, de las amas de casa y de las escuelas de niñas* (Bogotá: Imprenta de Gaitan, 1869).

21 Despite the lack of direct evidence able to present in detail the women's role in the creation of gardens has been a topic deeply addressed by literature studies from a whole range of epochs and geographies. In the particular case of Latin America, analysing the women' role in the Cuban society deployed in the Novel Sab, the scholar Naomi Lindstrom shows how in the nineteenth century Latin America, shows how urban like private gardens as well as those located in convents represent and attempt to create a space where they could enjoy of ways of satisfaction denied by the society, see Elizabeth A. Augspach, *The Garden as Woman's Space in Twelfth- and Thirteenth-Century Literature, Studies in Mediaeval Literature*, vol. 27 (Lewiston, NY and Lampeter: Edwin Mellen Press, 2004); Stephen Bending, "Melancholy Amusements: Women, Gardens, and the Depression of Spirits," *Studies in the Literary Imagination; Atlanta* 44, no. 2 (Fall 2011): 41–62, 133; Christine Coch, "The Woman in the Garden: (En)Gendering Pleasure in Late Elizabethan Poetry," *English Literary Renaissance* 39, no. 1 (January 1, 2009): 97–127, https://doi.org/10.1111/j.1475-6757.2009.01041.x; Stephanie Li, "Domestic Resistance: Gardening, Mothering, and Storytelling in Leslie Marmon Silko's Gardens in the Dunes," *Studies in American Indian Literatures* 21, no. 1 (2009): 18–37; Naomi Lindstrom, "El convento y el jardín: La búsqueda de espacios alternativos en Sab," *Decimónica 4.2* 4, no. 2 (2007): 49–59.

22 Mejía, *Los Años Del Cambio*.

23 Ibid.

24 Alberto González-García and Antonio Gómez Sal, "Private Urban Greenspaces or 'Patios' as a Key Element in the Urban Ecology of Tropical Central America," *Human Ecology* 36, no. 2 (2008): 291–300.

25 About the concept of infraknowledge in photography, see Roland Barthes, *Camera Lucida, Reflections on Photography* (New York: Hill and Wang, 1981).

26 These urbanistic regulations were issued by the kings Fernando el Católico in 1513, Carlos V in 1526, and Fernando II in 1573, see Silvia Arango, *Historia de La Arquitectura En Colombia* (Bogotá: Centro Editorial y Facultad de Artes, Universidad Nacional de Colombia, 1991), 87.

27 The idea of the city as a manifestation of the divinity that represents a 'psyche unity' is well presented by Joel Kotkin, see Joel Kotkin, *The City: A Global History* (London: Weidenfeld & Nicolson, 2005).

28 Mejía, *La Ciudad de Los Conquistadores 1536–1604*.

29 Pablo Rodríguez, "Casa y Orden Cotidiano En El Nuevo Reino de Granada, s. XVIII," in *Historia de La Vida Cotidiana En Colombia* (Bogotá: Grupo Editorial Norma, 1996), 103.

30 Martínez, "La Vida Material En Los Espacios Domésticos," 351.

31 Acosta de Samper, "Mi Madrina," 102.

32 Felipe Pérez, *Geografía General de Los Estados Unidos de Colombia y Geografía Particular de La Ciudad de Bogotá*, vol. I (Bogotá: Imprenta de Echavarria Hermanos, 1883), 16.

33 Rainer W. Bussmann et al., "Astonishing Diversity—the Medicinal Plant Markets of Bogotá, Colombia," *Journal of Ethnobiology and Ethnomedicine* 14 (2018), https://doi.org/10.1186/s13002-018-0241-8; Sara E. Giraldo et al., "Ethnopharmacological Studies in Traditional Markets from Bogotá D.C. (Colombia): Advances and Perspectives for the Search of Medicinal Plants with Therapeutic Potential," *FarmaJournal; Salamanca* 4, no. 1 (February 2019): 253–54.

34 Issac Holton, *New Granada. Twenty Months in the Andes* (New York: Harper & Brothers, 1857), 140.
35 Felipe Pérez, *Jeografia Fisica i Politica de Los Estados Unidos de Colombia*, vol. I (Bogotá: Imprenta de la Nación, 1862), 16, http://hdl.handle.net/2027/wu.89044216349.
36 Holton, *New Granada. Twenty Months in the Andes*, 140.
37 Acosta de Samper, "Mi Madrina," 22.
38 Pérez, *Geografía General de Los Estados Unidos de Colombia y Geografía Particular de La Ciudad de Bogotá*, 16–17; Holton, *New Granada. Twenty Months in the Andes*, 9.
39 Steuart, *Bogotá in 1836–7: Being a Narrative of an Expedition to the Capital of the New-Granada and a Residence There of Eleven Months*, 151.
40 Martínez, "La Vida Material En Los Espacios Domésticos," 351.
41 Ian Cook and Philip Crang, "The World On a Plate: Culinary Culture, Displacement and Geographical Knowledges," *Journal of Material Culture* 1, no. 2 (July 1, 1996): 131–53, https://doi.org/10.1177/135918359600100201; Germán Patino Ossa, *Fogón de Negros: Cocina y Cultura En Una Región Latinoamericana, Segunda edición, Biblioteca Básica de Cocinas Tradicionales de Colombia 8* (Bogotá: Ministerio de Cultura, 2012).
42 Mejía, *Los Años Del Cambio*.
43 Bussmann et al., "Astonishing Diversity—the Medicinal Plant Markets of Bogotá, Colombia"; Giraldo et al., "Ethnopharmacological Studies in Traditional Markets from Bogotá D.C. (Colombia)."
44 María C. Sierra-Guerrero and Angela R. Amarillo-Suárez, "Socioecological Features of Plant Diversity in Domestic Gardens in the City of Bogotá, Colombia," *Urban Forestry & Urban Greening* 28 (December 2017): 54–62, https://doi.org/10.1016/j.ufug.2017.09.015.

3 The 'invisible' flora

The flora of unattended commonplaces

Not all the plants in the nineteenth-century Bogotá were cultivated. The natural development of the city had left many residual spaces and unattended commonplaces: interstices, remaining spaces, dead zones, or urban voids.[1] These unurbanised and transit spaces were either paved or unpaved streets, abandoned houses and solares, undeveloped plots, meadows and paddocks, and riverbanks. Together, all of these spaces made the city a heterogeneous landscape where urban, rural, and wild realities were constantly interwoven and influenced each other (see Figure 3.1). These unattended and no-body spaces were rapidly colonised by spontaneous plants, and 'urbaniphilous species' adapted to the urbanised environments introduced during different periods of the city's history.[2] The nineteenth-century flora of open spaces within Bogotá made this city a 'botanical palimpsest' formed by native plants growing alongside introduced species, some of which, such as mustard (*Brassica rapa*), had naturalised a long time ago.[3] Therefore, in the no one's urban spaces [today and then], European weeds and spontaneous plants such as dandelion, greater plantain (*Plantago major*), clover (*Trifolium repens*) and chamomile, commonly grew side by side with the native curuba (*Passiflora tripartita*), *uchuva* and *guasca* (*Galinsoga parviflora*).

Transit and residual spaces in Bogotá were not only inhabited by 'weeds' and 'escaped' plants from patios and solares, but they also had trees and bushes. Species thriving in highly modified environments such as *aliso* (*Alnus acuminata*), or *cerezo* (black cherry, *Prunus serotina*) were common in the city's riverbanks, meadows or abandoned solares (see Figure 3.2) On the other hand, the high-water table of Bogotá's soil, plus the ability to grow from sticks, seems to favour the development of *sauce llorón* (*Salix humboldtiana*) which were abundant in the city. The Geographer Felipe Pérez when describing the general aspect of Bogotá in 1862 observed how 'thousands of willows adorned the city'.[4]

DOI: 10.4324/9781003208167-5

Figure 3.1 Ramón Torrez Méndez, *Boquerón del Río San Francisco* [See the rural-like environment of Bogotá] ca. 1837, oil on canvas, 63.2 × 103.5 cm.

Source: Courtesy of the Museo de la Independencia

Meanwhile, either driven by natural processes of dispersion or as a consequence of some people's will, isolated trees and large bushes were also part of the urban flora. This is evidenced in the historical names held by some streets in the city. Streets in Bogotá were named according to their features: 'A church, an effigy of a saint on a wall, someone's remarkable house was used in the official "baptism" of streets.'[5] Following this tradition, trees and plants, in general, were also commonly used as toponymical elements in the colonial and early Republican periods eras. Streets named after plants included *Calle del viñedo* (Grapevine Street), *Calle del Olivo* (Olive Street), *Calle del Curazao* (Bougainvillea Street), *Calle de la curuba* (Banana Passionfruit Street), *Calle del espino* (Thorn-plant Street), *Calle de la papaya* (Papaya Street), *Calle del nenufar* (Water Lily Street), and *Calle del Chocho* (Chayote Street).[6] As with today's barren trees in *Calle del Cedro* (Andean Cedar Street) in the centre of Bogotá, street names have outlived the physical existence of the plants on which behalf they had been named. The result of this was a lingering symbolic presence of plants in the modernised city whereby the trees and bushes present in Bogotá were memorialised as ghost flora.

Considering the historical lack of large-sized trees in Bogotá, as explained in the final section of this book, it is notable the existence of streets named after large size trees such as *cedro* or, moreover, the still existing *nogal*

Figure 3.2 Anonymous, *Plan of a solar by the Rafaela Velasco shop* [see the representation of some spontaneous and cultivated plants], 1800, ink in paper, 56 × 41 cm.

Source: Courtesy of the Archivo General de la Nación

(Colombian walnut), which gave its name to an entire neighbourhood in the north side of the city. Since most of the big pre-Columbian trees around the Muiscas' settlements disappeared with the construction of the European colonial city, it is possible that such trees would have been planted during the early colonial period. However, a lack of evidence prohibits an estimation of the number and density of trees growing in the residual and transit areas of the city.

During the last decades of the nineteenth century, Bogotá underwent an accelerated process of urbanisation. Thus, while houses of patios were flatted and the solares were actively used to build new edifications, streets were macadamised first and then paved, thus covering forever the soil where ruderal species used to grow.[7] Additionally, during the first decades of the twentieth-century rivers were domesticated and controlled to allow the rapid urbanisation of the city. The rivers San Francisco and San Agustín, among which the city had been founded in 1538, were corrected, canalised and later disappeared under a layer of cement. All the infrastructure transformations led to the destruction of river environments that had historically hosted different species of plants linked to these river currents. Thus, although many of the plants growing in early nineteenth-century Bogotá were invisible for most of the population, the work of the botanist José Jerónimo Triana gives us an idea of Bogotá's uncultivated flora. In the development of his unfinished work *Flora de Bogotá*, Triana collected different species occurring in and around including an *Arum italicum* growing by 1856 on the San Francisco River's bank. This plant, however, is not native to the Andes. *A. itlaicum* is a species native to Europe and used largely as an ornamental. Its presence in the urban environment of the mid-nineteenth-century Bogotá shows how some plants not necessarily edible or medicinal were naturalised in Bogotá. Additionally, this botanical collection raises questions about Triana's reason for collecting this unique plant ignoring, for instance, other natives that perhaps were quite common for him.

Wild plants outside the city

The invisible flora of Bogotá was everywhere. Plantago, dandelions and cloves grew in any possible place. In the meantime, *curubas, uchuvas* and mustards did so in any place where a little available soil. Some of these plants were potentially spotted by children, and some of their fruits were picked from people who found them on their way, but generally speaking, their existence was hidden by most of the people in the city.[8] However, there was an occult flora out of the sight of most of the people in the city. This was formed by various species of plants thriving wild around the city, some of which had a central role in the quotidian urban life. Through a series of selected examples, this section presents instances of how the city used plants from forests, swamps and *páramos* around Bogotá and how this extractive process was possible thanks to the vernacular botanical knowledge of the socially excluded people who were responsible for ensuring a steady supply of necessary raw materials for daily urban life functioning.

Non-timber forest products (NTFPs) as substitutes for industrial products

In many different ways, the nineteenth-century Bogotá was a city anchored in its past. Its cultural practices still reflected the 'ritual atmosphere' that came with the mediaeval ideas implanted by the Spanish during the conquest.[9] Contrary to the countries in the south of Latin America, Colombian industrialisation proceeded slowly and on a rather small scale. Although it had increased since independence in 1819, trade developed slowly due to precarious infrastructural conditions worsening the city's historical isolation.[10] Under these circumstances, many of the material elements consumed by people in Bogotá were extracted or produced locally and regionally using non-timber forest products (NTFPs) as raw materials. Therefore, these plants or products deriving from them were used in an incredible range of activities. In 1888, a journalist pointed out this situation as shown as follows:

> Many years will pass, centuries perhaps, before cultivated Europe gains sufficient knowledge about the richness, variety and luxury of the Colombian plants. Was right the person who declared to have been found in the vegetal realm almost everything produced by humans around the world. Plants of nails, wire and springs; of wires, cords and cables, of vessels. Strong cups and spoons, [plants] of all diameter tubes, of candles, of liquors, of tills, and know not what more else, produced by our generous nature to make bearable the isolation in which we live; and to compensate in part the difficulties that do not allow us to be in contact with the large civilization centres.[11]

Colombia's enormous biodiversity, relative geographic isolation and underdeveloped national industry presupposed the prevalence of the use and exploitation of products coming from the wild. The Bogotanians of the preindustrial city slept on mats made of *totora* (*Schoenoplectus californicus* subsp. totora), a semi-aquatic plant abundant in the *Tota* lagoon (20 km away from the city). Also used as carpets, these vegetal mats were used to provide adequate isolation against the chilly weather of a city located at 2,600 m a.s.l. in the Andean mountains (see Figure 3.3). Plants were also used to produce garments. For instance, the so-called *sombrero de Panamá*

Figure 3.3 Gumercindo Cuellar, *Campesinos* [mats made of totora from the Tota Lagoon], ca. 1930, photograph, 9 × 15 cm.

Source: Courtesy of the Biblioteca Luis Ángel Arango with authorisation of Mario Cuellar

(Panamá hat), a highly estimated piece of handcraft commonly worn by men, was made through the use of a vegetal fibre obtained from the *iraca* (*Carludovica palmata*), a middle elevation plant collected on the steep sides of Andes mountains. Similarly, people used the fruit of a gourd-like Asian plant called *estropajo* (sponge ground, *Luffa aegyptiaca*) as a sponge to clean bodies and dishes. Mops and brooms were also made through the use of plants. Likewise, bamboo sticks were used as tubes for water collection from public fountains in a city that long ignored the existence of hoses.

Plants were not only present in Bogotá's kitchens in the guise of vegetables used in the hybrid Colombian cuisine, but they were also broadly used as kitchen utensils, taking into account that the first pottery factory was not opened in Bogotá until 1833 and that there was not a single cutlery factory in Colombia during the first decades of the nineteenth century.[12] The most straightforward kitchen tools were derived from plants. Perhaps the most remarkable example is the use of *totumos* (calabash trees, *Crescentia cujete*), a low-land plant producing thick and durable fruit, in the production of spoons, plates, bottles, and cups. In a time in which glass was not a common material and plastic was far from being invented, *totumo* bottles were the most available vessel to carry and transport liquids. The traveller James Steuart left a description of the extensive use of *totumos* as cups and glasses:

> *Chicha* is in general use. This is a peculiarly the drink of the poor, although I have seen even foreigners use it. It is kept in huge earthen jars, wrapped round with green hide to preserve it; it sells for about four or six cents a quarts. A large *tituma* (sic) is filled, which is passed round from one to another throughout even a large company![13]

He also highlights how:

> The coffee and the chocolate are not drunk with their meat, but immediately afterwards, and then the same small cup is made use of in the morning. The disgusting habit of a whole family using one red earthen goblet or a *tituma* (sic) in common for drinking from is universally prevalent.

Whereas *totumos* were used to transport small quantities of liquid, woven baskets were largely implemented for carrying various kinds of solid products. Much like modern plastic bags, woven baskets made of fibres or vine shoots were broadly used by the Bogotanians. However, their most important role was probably at the market. Baskets were used as containers in which people from the countryside brought their merchandise to be sold at the Friday market in the central city square. For instance, it was common to see people transporting hens or chickens in small fibre-made cages specially made for poultry (see Figure 3.5 right). Similarly, people who transported other NTFPs used woven bags to bring their herbs or tree barks to Bogotá. Moreover, baskets and vegetal woven bags were profusely sold in the market. With no other elements on hand to transport goods, these baskets and vegetal bags were quotidianly used by people carrying home the groceries they bought in the market.

One of the most important uses of plants was in the physical construction of the city. To create a Spanish-like city, the conquerors had to take advantage of what was at hand. This meant to employ a native workforce:

people embodying vernacular building practices, which included a knowledge set of the natural resources involved in construction. Consequently, although the colonial city was eminently a European project, the truth is that it had to be developed using local people's building knowledge, which largely involved a deep botanical knowledge of plants used as construction raw material.

Beyond the evident use of timber, colonial building construction demanded significant amounts of other types of vegetal raw materials. The bamboo-like *chusque* (*Chusquea scandens*) and the *pajonal* (*Calamagrostis effusa*) played prominent roles in the construction process. From the latter, the Muiscas produced a highly resistant braid known as *cuan* used to join *chusque* rods and, in doing so, roof both natives' and conquerors' houses (see Figure 3.4).[14] The skilled use of straw (probably of *C. effusa*) in conjunction with chusque was early recorded by the Europeans when they arrived in Muisca's lands. In 1550, the Spanish found a settlement which they called Pueblo Nuevo [today's Cajicá]. Describe it as 'very beautiful with few but big finely thatched houses, which were well fenced with rows of rods gently put together with art'.[15] The ethnobotanical knowledge involved in the

Figure 3.4 *Chusque* and *pajonal* cord (*cuan*) extracted during the refurbishing of the sixteenth-century convent of Santo Domingo, Tunja, 2017 digital photograph.

Source: Author's image

construction skills not only survived the conquest but was essential in the construction of the colonial and even the nineteenth-century republic city.

Although economic constraints imposed severe limitations on the urbanisation process, due to their physical resistance and durability, the *chusque-cuan* couple was used each time a new house or public building was built. Thus, the transplantation of the colonial house originating in Spain, which was inhabited mainly by Spanish or Spanish descendants [known as *criollos*], was made possible thanks to the accumulated botanical and architectonical knowledge used in the construction of indigenous people's spaces.

The outstanding physical characteristics of *chusque-cuan* roofs also enabled this building technique to be integrated into the city's material transformation under modernisation, as evidenced by the presence of *cuan* and *chusque* in a construction list published in the *Anales de Ingenieria* (Engineering Annals) in 1887. The list reveals that among many other construction raw materials such as timber or nails, *chusque* was sold in parcels of 25 pairs and a package of 25 pairs of *cuan* fibre was sold for 25 *pesos*.[16] Some years later, when Tomas Cipriano de Mosquera's government seized the Catholic Church's assets in 1906, he planned to build a new armoury inside the San Agustin Monastery. As part of the requirements to accomplish the building, the engineer, Mr Jaramillo, requested 880 packages of *chusque* and the same quantity of *cuan*. Each of those packages was composed of 25 pairs of chusque sticks and 25 caun braids; thus, the overall number of those two vegetal products required for the armoury construction was around 20,000! This figure suggests a steady high level of exploitation of *chusque* and *pajonal*, therefore implying an intensive alteration of their natural populations, which likely enormously affected the *páramo* and cloud-forest ecosystems where *chusque* and *pajonal*, respectively, develop.[17]

Despite the high quality of the native materials used in the construction of Bogotá, the consolidation of modern ideas brought about scorn for the city's Hispanic heritage and the consequential abandonment of colonial ways of building, and the French-like paradigm became the model of progress. Modern architecture meant not only new building styles but also the use of new materials such as cement, which was first introduced in 1881, and perceived as one of the traits of progress.[18] The technological transformation during the last decades of the nineteenth century would forever replace the autochthonous techniques of the pre-Columbian vernacular architecture, at least in the homes and buildings of the ruling classes and other elites. Thus, along with the desperate attempts to erase the colonial past emerged another contradiction: the end of the conqueror era meant the erosion of some of the Muiscas' cultural traditions and knowledge that had previously survived among the upper and middle classes.

However, the modernising French architecture influx did not spread among all Bogotanians. The historically marginalised native American population continued applying specific traditional botanical and architectural knowledge in the construction of their houses. After centuries of spatial segregation in the *pueblos de Indios*, the Muiscas' descendants gradually moved into Bogotá, carrying their building practices with them. Lacking the resources (or willingness) to build their own houses according to the modern styles, they applied vernacular techniques and vegetal raw materials such as *chusque* and straws to create their own homes. For instance, they used *chusque* as a sort of structural skeleton in the construction of mud walls in the so-called *casas de bahareque*. Straw, coming from the same *pajonal plant* or wheat plantation leftovers, was not used in the fabrication of *cuan* braids; instead, it was devoted to roof construction. Overall, this process produced the *casas pajiza* or *rachos* (thatched huts) that have been historically synonymous with poverty. The persistence of thatched houses throughout the nineteenth- and twentieth-century Bogotá was a reflection of the overextended poor living conditions of the city. Thus, while in 1878, the city's record registered 164 thatched huts; by 1907, there were 2,336 *ranchos* in the city.[19] The increase in these kinds of houses reflects how the growth of the city was based on the migration and reproduction of the poorest. These figures indirectly speak about a hidden pressure on the nearby ecosystems from where chusque and straw came from and also showed how through the manipulation of these plants, the poorest help to create their own marginalised space in the outskirts of the city through ancient ethnobotanical knowledge.

People in contact with wild plants in Bogotá were the poorest (see Figure 3.5). They had no access to places like patios decorated through the manipulation of ornamental plants, and only some of them were employed as servants, therefore having a constant interaction with plants cultivated in solares. Conversely, when they did not occupy the thatched houses, they lived in small, dark, and poorly ventilated rooms, commonly called tiendas. The botanist Isaac Holton described one of these rooms as 'The little room looks like a prison cell, only it has no grated window, nor loop-hole, nor-breathing-hole, except the open door.'[20] Without green domestic spaces, raw material sellers developed particular relationships with plants in the nearby Andean ecosystems out of sight from most people. Thus, if we think about the city's nineteenth-century city as a floristic island in the process of consolidation, people bringing NTFS from nearby ecosystems can be considered as kind of fishermen and fisherwomen. An example of this is the city's constant demand for firewood and coal supplied by the work of poor people. In the specific case of charcoal, they needed to walk long distances to find a suitable tree(s), turn it into charcoal, and afterwards carry

Figure 3.5 Left (Anonymous, Straw seller-woman in the nineteenth-century Bogotá, ca. 1890 photograph, 13.8 x 9.6 cm. Right) Anonymous, Men with chicken baskets, ca. 1890, photograph, 13.8 × 9.6 cm.

Source: Dubail, C., and M. Dubai-Acero, eds. *Bourgarel, le Colombien: Voyages d'un diplomate français dans la Colombie du XIXe siècle*. Paris: EdiSens, 2017. Courtesy of the Association Ernest Bourgarel

it down to the city to sell. Eugenio Díaz Castro, in his 1880 book *Bruna la Carbonera* (*Bruna, the Charcoal-maker Woman*), wrote:

> Toward the East, on the top of the Monserrate and Guadalupe mountains, there is a large extension of land covered by scrublands. From there, the charcoal and firewood sellers obtain their subsistence products that supply of fuel to the city. The vegetation of those mountains, barren at first glance, diminishes gradually until the hills are covered by "pajonales" [*Calamagrostis effusa*] and the "frailejón" [*Espeletia* spp.]. This is the end of the vegetal realm.[21]

As shown in this extract of literary narrative, the quote presents the harsh conditions faced by those who were in charge of supplying plants and NTFPs to the city. Equally, the iconographic evidence confirms how most of these people were part of the poorest class (see Figure 3.5). Socially and economically marginalised, some of the Muiscas descendants used the traditional ethnobotanical knowledge that enabled them to know, for example, the perfect moment to pick the 'estropajos' and 'totumos' from the plants, what kinds of vines were suitable for the construction of baskets, select the trees suitable to obtain charcoal, or to understand the best moment to extract from the nearby forest certain elements, such as moss and straw. Despite its importance, this preindustrial traditional botanical knowledge was fundamental for the functioning of the colonial and preindustrial city and has received no attention from scholars.

Notes

1 Claudio Curzio, "El Origen y Las Características de Los Fragmentos Urbano-Público Residuales," *Cuadernos Geográficos*, no. 42 (January 1, 2008): 53–82.

2 Marcus Nyman, "Drying Greens, Commons, and the Possibilities of the Botanical (in One Quiet Corner of London)," in *The Botanical City*, ed. Matthew Gandy and Sandra Jasper (Berlin: Jovis, 2020), 107–19; Rüdiger Wittig, Dagmar Diesing, and Michael Gödde, "Urbanophob—Urbanoneutral—Urbanophil Das Verhalten Der Arten Gegenüber Dem Lebensraum Stadt," *Flora* 177, no. 5 (January 1, 1985): 265–82, https://doi.org/10.1016/S0367-2530(17)30180-9; Rüdiger Wittig and Ute Becker, "The Spontaneous Flora around Street Trees in Cities—A Striking Example for the Worldwide Homogenization of the Flora of Urban Habitats," *Flora—Morphology, Distribution, Functional Ecology of Plants* 205, no. 10 (October 1, 2010): 704–9, https://doi.org/10.1016/j.flora.2009.09.001.

3 On the use of vegetation as a testimony of the city's history, see Moritz von der Lippe, "Vegetation as Testimony: Botanical Traces of Urban Past," in *The Botanical City*, ed. Matthew Gandy and Sandra Jasper (Berlin: Jovis, 2020), 46–53.

4 Pérez, *Jeografia Fisica i Politica de Los Estados Unidos de Colombia*, 11.

5 Eduardo Posada, *Narraciones*, 2nd ed. (Bogotá: Villegas Editores, 1988), 73.

6 Ibid., 73–74.

7 Mejía, *Los Años Del Cambio*, 38.

8 Plant blindness as explained by James Wandersee and Elisabeth Schussler is not only a contemporary phenomenon, it has historical roots that can be tracked, see The plant blindness James H. Wandersee and Elisabeth E. Schussler, "Preventing Plant Blindness," *The American Biology Teacher* 61, no. 2 (1999): 82–86, https://doi.org/10.2307/4450624.

9 Romero, *Latinoamérica Las Ciudades y Las Ideas*, 21–68.

10 Although Bogotá's isolation has been widely accepted by historians, its confinement was rather relative. On the one hand, Bogotá was the centre of a huge network of paths that linked the Muisca's territories with other regions; hence, the city became isolated after the conquest campaign when the local world was expanded and the administrative centre was located in Madrid. On the other hand, the poor physical conditions of trade paths and resulting long travel and communication times did not prevent the establishment of connections between Bogotá and the rest of the Western world, see Langebaek, *Mercados, Poblamiento e Integración Étnica Entre Los Muiscas Siglo XVI*.

11 Ramón Guerra, "La Guadua," *Papel Periódico Ilustrado*, October 1, 1888.
12 Martínez, "La Vida Material En Los Espacios Domésticos," 350.
13 Steuart, *Bogotá in 1836–7: Being a Narrative of an Expedition to the Capital of the New-Granada and a Residence There of Eleven Months*, 152–53.
14 *Cuan* is the Spanish translation of the muisca word *pquamne* that literally means *cord of straw*, see http://muysca.cubun.org/MU/cuan.
15 The original versión is as follows: *Era muy hermoso, de pocas casas y muy grandes, de paja muy bien labrada, las cuales casas estaban muy bien cercadas de una cerca de hazes (sic) de cañas por muy gentil arte obradas* Hermes Tovar, "relación de Santa Marta," in *Relaciones y visitas a los Andes, S XVI, Región del Caribe*, vol. Tomo II (Bogotá: Biblioteca Nacional, Colcultura, Instituto Colombiano de Cultura Hispánica, [ca. 1550] 1992), 170.
16 Julián Lombana and Francisco Ayala, "Precios de Los Materiales de Construcción En La Ciudad de Bogotá," *Anales de Ingeniería* 1, no. 1 (Agosto 1887): 32.
17 Arsenal y parque que se proyectó construir en el edificio de San Agustín, 20 Octubre 1906. Tomo 857 (parques, plazas y jardines), folio 285–7, Sección República, Fondo Ministerio de Obras Públicas. AGN.
18 Luis González, *Del Alarife al Arquitecto: El Saber Hacer y El Pensar La Arquitectura En Colombia: 1847–1936* (Medellín: OjoXojO, 2013), 172–79.
19 Germán Mejía Pavony, "Los itinerarios de la trasformación urbana Bogotá, 1820–1910," *Anuario Colombiano de Historia Social y de la Cultura*, no. 24 (January 1, 1997): 101–37.
20 Holton, *New Granada. Twenty Months in the Andes*.
21 Eugenio Díaz, *Novelas y Cuadros de Cosmtumbre: Bruna La Carbonera*, vol. II, III vols. (Bogotá: Procultura, [1880] 1985).

Section II

The creation of the urban nature

4 From colonial squares to hygienic gardens

Sanitising with plants

When in 1890 trees started to be planted in the city of Medellín, the writer Tomas Carrasquilla observed that: 'Planting without harvesting is an evident manifestation of urban progress, urban arborisation does not mean anything for the uncultivated peasant'. For Carrasquilla, planting trees in the city was the 'discretion of cultured people who do not live in the mountains'.[1] One of these urbanites mentioned by Carrasquilla was Genaro Valderrama. As the General Inspector of Parks and Public Gardens of Bogotá, he opined that cleaning houses and streets was not enough to keep the healthiness of the city if it was not combined with massive urban afforestation. According to Valderrama, since vegetation was 'the most potent agent that works upon the public health', the city needed more efforts to increase the poor vegetation existing in parks and public and private gardens.[2]

The nineteenth-century Bogotá was filthy. The luring light of modern innovations had promoted a massive migration of peasants ready to abandon their shovel and muddy lives to embrace the routine in the factory. Bogotá's population grew four-fold between 1851 and 1912. However, contrary to what happened in other capitals such as Rio de Janeiro or Buenos Aires,[3] the economic constraints inherited from the colonial era forestalled an urbanisation process that could properly host the city's growing population. To partially solve the problem, the last decades of the century experienced a large process of urban densification. Solares with their cultivated and spontaneous plants were urbanised while colonial houses of patios were subdivided to make space for the constant flow of people.[4] However, the concentration of people in more or less the same space as the colonial city was not enough to host the growing population (see Figure 4.1).

The housing problems of Bogotá were severe to the point of endangerment. For example, in 1890, the city had 10,000 buildings, 900 thatched huts and a population of 100,000 inhabitants. Those figures suggest that,

DOI: 10.4324/9781003208167-7

Figure 4.1 Ernest Bourgarel, *Street of Bogotá*, ca. 1895, photograph.

Source: Dubail, C., and M. Dubai-Acero, eds. *Bourgarel, le Colombien: Voyages d'un* diplomate *français dans la Colombie du XIXe siècle*. Paris: EdiSens, 2017. Courtesy of the Association Ernest Bourgarel

on average, each of those buildings was inhabited by at least nine people.[5] Overcrowding of this magnitude in a city with poor infrastructure in terms of sewers and without a proper water supply resulted in a constant presence of tuberculosis, typhus, smallpox, measles, and dysentery, as well as other contagious and non-contagious diseases.[6] The poor health conditions in Bogotá were reflected in the mortality rate; in 1865, 1,074 people were born in the city; however, 1,078 people died in the same year. Despite the high mortality rate, a steady migration process led to an overall increase in Bogotá's population during the nineteenth century.[7] Consequently, although its modern innovations such as electric energy and theatres made Bogotá the brightest and most alluring city in the country, its poor hygienic conditions and underdeveloped infrastructure ultimately made it little more than a *moridero* [a place to go and die].

The mismatch between the city and its population was evident on the streets. What has been transit and residual spaces colonised by plants adapted to urban environments, by the end of the nineteenth century, had turned into cesspits. In 1899 the French traveller Jorge Brisson observed that '[t]he streets are generally in an unbearable state, bad stone-paved, dirty most of the times, plenty of holes, mud and filthy'.[8] Added to this, free-roaming animals and those used to transport Bogotá's wealthy residents in carriages and similar conveyances filled the streets with excrement and urine.[9] However, what seems to have contributed most to the low levels of cleanliness of the streets was the lack of a proper sewer and drain system. Due to the lack of any other place to deposit their human faeces, streets were used as public latrines,[10] as was described by Sir James Steuart in 1836 when writing about the waters from the mountains that passed across the city:

> the smaller streams [of the city], which run down from the fountain-heads, and merely follow the indented pavement in the centre of the streets through which they flow, are an immense value to the citizens, as they empty into them every kind of filth and offal; the poor make their toilet by them.[11]

As underscored by Steuart, the poorest were seen as a source of filth. This is not because they inhabited *ranchos* and *tiendas* without latrines, but because they were many. The city's population growth was accompanied by a steady increase in social exclusion. Bogotá became the destination for many coming from nearby regions, as people uprooted by post-independence wars quickly moved towards the city. However, the lack of industry and other economic activities meant that the migrant population struggled to find adequate living standards and generally faced harsh conditions.[12] This increase in poverty did not go unnoticed by the emergent elites. The local press frequently published articles denouncing the presence of herds of 'immoral' beggars, ill people, orphans, and thieves on the city's streets. In a seminal intellectual effort to understand the origins of poverty in the city, Colombian lawyer and politician Miguel Samper, a member of the Liberal Party, published a small book titled *La Miseria en Bogotá* (1867), in which he decried how 'Streets and squares of the city are infested by thieves, drunks, vagabonds, lepers and even lunatics'.[13] The now evident increase in poor people was perceived by the traditional inhabitants of

Bogotá as a cause of shame, and as a problem that must be solved in order to achieve the civilised status associated with the modern lifestyle.

Because the link between moral decay and physical illness was common during the nineteenth century,[14] physicians were responsible for solving the emergent urban social and environmental issues, alleviating the overcrowded, ill, and 'degenerate' city. Reconsidered as the early planners of the industrial city.[15] doctors' opinions spanned a range of urban topics. For instance, in 1886, the chief of police sent a letter to the *Sociedad de Medicina y Ciencias naturales* (Medical and Natural Science Society) asking for guidelines to improve the moral and health conditions of the city. In response, this Society carried out a series of investigative commissions to devise solutions to many of the city's problems, including public water (latrines, river, sources, and distribution of drinking water); street cleaning, including the collection of rubbish and dead animals; hygienic conditions within industrial and trade places such as pharmacies, taverns, bakeries, breweries, or stables; pornography; education; and the regulation of public spaces including marketplaces, cemeteries, hospitals, orphanages, jails, and *parks, gardens, and public promenades*.[16]

Considering the hygienic role of plants as 'lungs' capable of ameliorating the filthy conditions of Bogotá, doctors prescribed plants to the ill city.[17] Doctors were responsible for administrating the initial efforts to use trees in cities. In many cases, they played an active role in suggesting the most suitable places and species for tree planting in the city. For example, in 1889, the Medellín-based physicians Francisco Arango and Francisco Mejía proposed that trees should be planted near the city cemetery to tackle the diffusion of gases toward inhabited areas.[18] Furthermore, the beautifying power of plants was highlighted alongside their health benefits as a part of the arguments presented by doctors in defence of urban trees. For example, in 1888, Francisco Uribe, another physician from Medellín, argued that hygiene 'is not only related to the maintenance of the healthy conditions, but it must address the physical and moral improvements and to embellish our surroundings used as public ornament'.[19] The aesthetic dimension of plants, namely their ornamental characteristics, was therefore assimilated by hygienist ideas that influenced the layout of the modern city. In some instances, including in Bogotá, the hygienic aspect of plants was subordinate to their ornamental use as instruments of embellishment. For instance, while in the process of transforming La Plaza de las Nieves (Nieves Square) into a public garden in 1889, Sebastian Castell from the police department sent a letter to the Minister of Economic Affairs in which he recommended that

> Knowing that the Honourable Member receives with his characteristic kindness any suggestion looking for the city improvements. I allow

> myself to highlight that the Las Nieves Square, because of its current condition is suitable to make of it an adorn of the city.[20]

Therefore, urban spaces endowed with plants were not only used as an organic filter to clean the air, but they also served as a raw material to embellish the places through which urban life passed.

In addition to its role as a consultant body, the medical guild endeavoured to spread the virtues of having plants as part of the city. Through their mass media magazine, the *Revista Médica de Bogotá* (Medical Journal of Bogotá), they regularly extracted sections of books published overseas addressing the importance of plants in cities. One telling example of this is the reproduction of the book entitled *Higiene Pública y Privada* (Public and Private Hygiene) published by the Spanish physician Francisco Javier Santero in 1886.[21] Santero remarked upon the hygienic properties of plants in urban environments:

> Squares and promenades fill a real hygienic need. They are true chimney of breath in relation to the streets that end in them and, therefore, they create air currents that purify the atmosphere. . . . Promenades must be formed by trees and not exposed to either strong airs nor close to points that emanate miasmas or unpleasant smells. Those [promenades] must be multiplied in the centre and the urban borders. . . . Trees influence the air composition given to it humidity, oxygen and ozone. However, the real importance of urban trees is their role as drainage through their roots.[22]

Santero's opinions do not only reflect the hygienic dimension of trees, but he also provided guidelines for planting trees on the streets. For example, he advises that 'streets with less than twenty-five meter-wide must not have trees' and recommends that planters 'take care that the tree's branches head the centre of the street and never toward houses'. In his paper, Santero also talked about plant species. He mentioned, for instance, a list of 14 suitable species of trees to be planted on the streets. However, the trees were European species suitable for Spanish cities and not for an Andean tropical city such as Bogotá. Interestingly, this fact did not promote any discussion among the local physician involved in the publication of the Medical Journal of Bogotá.[23] In any case, Santero's guidelines for structuring urban nature reveal how in addition to city planners, doctors could be considered among the first modern urban landscape designers and the first experts in urban forestry.

Partially because of its role as hygienic device but also because having gardens was seen as a symbol of progress, during the last decades of the nineteenth-century parks and modern promenades were gradually created

Figure 4.2 Anonymous, Avenida de la República-Bogotá [Today's Avenida Jiménez], 1910, photograph, paper on polyester 9 × 29 cm.

Source: Courtesy of the Biblioteca Publica Piloto de Medellín

in Bogotá (see Figure 4.2). Slowly, in a small town located 2,600 m a.s.l., plants that had previously been confined in patios and solares leapt into public spaces and helped to recreate the urban green spaces required for any city that wanted to be perceived as modern.

Creating public gardens

One of the characteristics of the modern city was its turn towards public life.[24] People's relationships guided by capitalistic dynamics move outside the domestic realm and plants follow the same direction. During the last decades of the nineteenth century, the need for public green spaces was pressing. The hygienic conditions of the city were lamentable, and an increasing mass of poor, ills and vagabonds cried out for elms. In the meantime, an incipient but growing labour class needed spaces for leisure that, in the eyes of the elites could keep them away from booze. For these elites, the lack of urban spaces ornamented with plants was a perceived of backwardness; hence, they actively work in creating a public garden. However, the creation of a public green spaces was not straightforward. To emulate the big projects developed by Georges-Eugène Haussmann in Paris was far beyond the possibilities of Bogotá's tiny budget. Additionally, the endemic lack of resources had forced the capitalist modernisation of Bogotá to squeeze into its colonial-built environment. Emergent ways of urban life and facilities such as the opera or the tram were imbibed in the colonial shell and the public nature did the same. Thus, the first public garden of Bogotá was established in the symbolic heart of the colonial and republican city: La Plaza Mayor (the main square).[25]

La Plaza de Bolívar

In 1880, the country's Minister de Education designated the self-taught gardener Casiano Salcedo to build a park or garden in the Plaza de Bolívar

(Bolívar Square).[26] Originally a significant Muisca astronomical site from which to observe the solstices and equinoxes,[27] established as the *Plaza Mayor* by the Spanish, re-named in 1821 as *Plaza de la Constitución* (Constitution Square), and finally turned into Plaza de Bolívar in 1846, this square was surrounded by significant buildings that have long served as the symbolic centre of social activity and performance in Bogotá.[28]

The decision to transform the most iconic site of the city into a garden reflected an entirely new approach to urban public space. This physical shift in the organisation of the town's main square was not a minor event. For the Spanish Philosopher, José Ortega y Gasset, the main square is the quintessential of the western urbanism. For him, the square is the negation of the rural life and its vegetation, it is an active 'separation' from nature executed by Greeks and Romans to discuss the *Res publica*.[29] However, different to Europe, where the middle age has spread religious and civil buildings across the cities, the Latin American city recovered the square as the city's political heart. As, Fabio Zambrano, one of Latin America's preeminent urban historians explained:

> In Latin America the main square was synonymous with municipality. It was the place where the *rollo* [a column stone which highlighted the administrative independence category of the city] and the pillory were placed. [The main square] means the starting point of the foundation act and the place where the symbol of justice rested. The importance of the main square is such that it represents the Hispano-American model of urbanization. Its use, loaded with symbolic meanings, defines the image of the city and express the idea of centralism present since the very beginning in this way to understand the city.[30]

The symbolic importance of the Plaza de Bolívar made it a gathering point where essential parts of the city's daily life were represented. For instance, religious processions and Republican celebrations took place in this square, and it was also the venue for criminal executions, public speeches, and bullfighting, among other activities. Bogotá main's church, known as *La Cathedral*, was also located in the main square. Its atrium, called *El Altozano* by the Bogotanians, was the most important meeting site for most of the population. The forever turbulent issues of political life were discussed at the Altosano. There, people shared news from Europe. Also there, while some educated people discussed literature or fashion, the young men gathered to flirt with ladies as they departed from the church. As described by Miguel Cane, this was the social heart of the city:

> You could ask, perhaps the Bogotanians do not use to promenade, they do not have a gathering point such as a club, a predilect street or something like our Florida street, the Corso of Rome, The New York's

Broadway or the London's Park Corner? I would say, yes, but everything in one: they have El Altozano' (see Figure 4.3).[31]

Meetings in the Plaza Bolívar acquired a special dimension on Fridays when the weekly market took place. From early in the morning, those who commonly frequented *El Altozano* mixed with the hundreds of people coming from neighbouring towns, who filled the square selling an incredible number of products. The vivacity and the multi-coloured nature of Bogotá's market overwhelmed foreigners. For example, Édouard André, the French Botanist visiting Bogotá in 1876, described seeing red, yellow, and white *ocas* (*Oxalis tuberosa*, ovoid cabbages, small white thistles, maize, red-necked

Figure 4.3 Anonymous, Plaza de Bolívar, ca. 1870, photograph.

Source: Courtesy of the Sociedad de Mejoras y Ornato de Bogotá green onions *(Polymnia edulis), arracachas* (*Arracacia xanthorrhiza*), garlics [*Allium sativum*], green peas [*Pisum sativum*], brussels sprouts, broad beans [*Vicia faba*], snow peas, chicory, and beans. He also saw crabs and fish, different kinds of poultry, cheese, milk, and an incredible number of fruits, including banana passionfruit (*Passiflora tarminiana*), Peruvian groundcherry (*Physalis peruviana*), *chulupa* (*Passiflora maliformis*?), papaya, and berries of different types. In addition to enumerating the products for sale at the market, André also portrayed the social dynamics that occurred each Friday in the main square. For instance, he mentioned how at the end of each Friday, *chicha*, *guarapo*, and *aguardiente* [all alcoholic beverages] sellers began to supply their 'toxic' product, the result of which was drunk peasants provoking several fights and arguments in public space.[32]

The most straightforward consequence of these kinds of social activities was filth. According to the hygienic-based ideas of doctors, markets were seen as unhealthy places that needed to be regulated. Attempts had been made to deal with this issue since the early-nineteenth century; eventually, proposals were developed to create a new roofed market that would engender healthier conditions.[33] In 1861, the Federal District Governor created a bill to move the Plaza Bolívar's market to the Santander, San Victorino, and San Agustín squares.[34] This idea was soon followed by the first proposal to build a public garden in the square. The National General Archive has preserved an outline of an English-style garden to be built in the Plaza Bolívar that was apparently created by a British trader named Jorge Bunch in 1861 (see Figure 4.4);[35] however, the design was never implemented. The eventual garden's construction commenced around 20 years later, when under the direction of Casiano Salcedo, a French-inspired design based on oval flowerbeds was finally adopted in 1881 (see Figure 4.5).

As I will explain in depth later, the creation of a garden in the political, social and economic centre of the city implied a transformation in the way of understanding and interacting with plants in the city. Also, having a garden in the main square meant a radical change in people's daily life with concomitant impacts on the ways that Bogotanians socialised. The garden consequences in the lives of ordinary people such as vendors or market customers are difficult to trace; however, newspapers documented the discomfort of some elite members. For example, a note published in a pro-Conservative Party newspaper bemoaned the inconvenience of having a garden surrounding such an important national hero:

> It is quite ridiculous to surround a hero statue with flowers, [This is] perhaps a sign of the excessive adulation that intoxicates and spoils them. But, doing that with Bolívar in the main square, would obscure such a delightful piece of art!'[36]

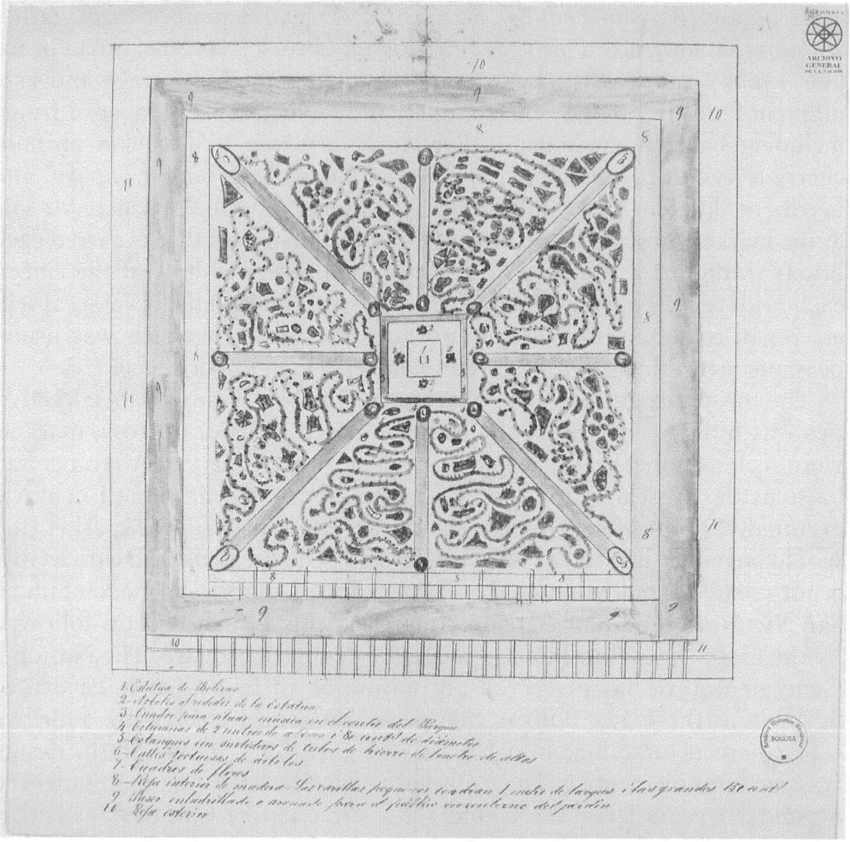

Figure 4.4 Jorge Bunch, *Ornamentation plan for the Plaza de Bolívar* [first design for an English-style garden for Plaza de Bolívar], 1861, watercolour in paper 33 × 33 cm.

Source: Courtesy of the Archivo General de la Nación

Such comments reveal the political connotations of transforming the Plaza de Bolívar into a garden. Other criticisms published in the same newspaper indicate that the garden enterprise was perceived by some elite sectors as a clear improvisation that evidenced a lack of formal horticultural knowledge. As one commentary acidly noted:

> Is hard to believe that cultivated and intelligent people help [in the construction] of such a difficult to define thing they are making around the statue. This riddle cannot be a garden because in such wide beds it is impossible to grow and take care of flowers. Not a grove since trees require circular beds. It is not a promenade, because its narrow paths do

Figure 4.5 Julio Racines, Garden in the Plaza de Bolívar in Bogotá, ca. 1890, photograph, 13 × 18 cm.

Source: Courtesy of the Biblioteca Pública Piloto de Medellín

> not allow it. It is not a farmyard because these do not require imported fertilized and rather produce it. It is not a kitchen garden, unless of official *arracachas* (*Arracacia xanthorrhiza*) that would be excessively expensive.[37]

However, despite the enormous transformation imposed on daily lives and the opposition of some members of the elites, the reshaping of the colonial squares continued and soon transformed other iconic spaces in Bogotá such as the Parque Santander.

El Parque Santander

In 1877, the Cundinamarca Government decided to flatten El Humilladero, the city's oldest chapel (see Figure 4.6). The demolition of this small building located in the Plaza de Santander was accomplished as one of the first attempts to create a public garden in the Plazuela de San Francisco. Located at the northern edge of the city on the way to the city of Tunja, the site had

Figure 4.6 Demetrio Paredes, *La Plaza* [Plaza San Francisco later renamed as Plaza Santander], 1868, photograph.

Source: Courtesy of the Leibniz Institute for Regional Geography, Germany been used in pre-Colombian times as a gathering place for intra-regional trade between the Muisca and the Zaque societies.[38] Given its strategic location and despite the official relocation of the city market to the Plaza Mayor, the square continued to informally and spontaneously serve as Bogotá's main market through the colonial period. In fact, it was due to this role that the square, which had been officially named Plazuela de San Francisco in honour of Francis of Assisi, was more commonly referred to as Plaza de las Hierbas (Herbs Square). As in the case of the Friday market at the Plaza Mayor, the natural dynamic of selling and buying live animals, vegetables, meats, and many kinds of non-timber forest products (NTFPs), resulted in notoriously filthy conditions. Added to this, because of its fountain, the Plaza de Santander was a gathering point for a considerable number of *aguateros* (water carriers) which, given the usually short supply of water, produced constant fighting and public disorder.

After the Humilladero chapel was demolished, the fountain was replaced in 1878 with a European-made statue of the national hero Francisco de Paula Santander. Taking the figure of Santander as a focal point, the square began to be decorated with plants. However, this plant-based ornamentation was rather a naïve enterprise. Due to the lack of a formal gardening culture in the city, the plants used to decorate the square were gathered from the population through a donation system. For example, a pine tree donated and planted by a man named Elias Garay remained in the square for many years.[39] Although this public-derived ornamentation can be counted among the early attempts to create gardened public spaces in Bogotá, it was not until 1880 that a systematic gardening process began, when the Plaza de Bolívar gardening project in turn promoted the transformation of the Plaza de San Francisco into a garden and led to its renaming as the Plaza Santander.

As part of the program to modernise urban spaces in Bogotá, Casiano Salcedo began to transform the Plaza Santander into a park in 1881. As described by Lisímaco Palau,

> the complete area was grassed, the park was divided into two parts, one with flowerbeds and the other with a group of trees. Additionally, two bronze fountains, wooden benches and a surrounding iron fence with a western and an eastern entrance were installed.[40]

The creation of the fenced park in Santander Square displaced the local market; however, even as late as 1894, people continued selling goods in one of its corners.[41] The spatial transformations led to a rather empty appearance of the brand-new parks at most times, as illustrated in Figure 4.7. The evidence indicated by the photographic sources is reinforced through some contemporary comments about the park. For instance, the Spanish traveller, Miguel Cané, wrote in 1881: 'Unfortunately, this spot [the Parque Santander] that could be a nice place for meeting, is generally a desert as occurred with the Calle de las Nieves and the Plazuela de San Diego'.[42] Nevertheless, the space previously occupied by the market gave rise to the emergence of new social dynamics with the establishment of weekly night-time parties (*retretas*) held each Thursday and Sunday.[43] Changing ways of social gathering brought new patterns of social behaviour, and differentiation of the social classes was performed as upper-class Bogotanians represented ostentation and manners imported from Europe along with the ideas of parks and public gardens. In short, the transformation of the square into a garden made it a site of modern behaviours and replaced the formerly spontaneous manifestation of human relationships.

The markets organised in the squares later transformed into parks were, in the eyes of Édouard André: 'The photography of a life, [where] the different social classes expose themselves naked, *without pose*, defending in public a way of existence that Darwin did not write about.'[44] In contrast, as I will discuss in Chapter 7, parks and public gardens imposed regulation on citizens' bodies, thereby establishing a more disciplined way of being. Any common people who wanted to breathe the pure air emanating

Figure 4.7 Anonymous, Parque Santander, ca. 1910, photograph (negative), 4.4 × 6.7 cm.

Source: Courtesy of the Museo de Bogotá

from the plants in the new parks and public gardens would need to behave themselves, ensuring that they did not perturb the enforced calm of these places. Finally, it is very interesting how accurately certain types of plants in the market and then in the garden reflected the transformation that had occurred with the modernisation of Bogotá. If the markets were 'inhabited' by edible plants cultivated by farmers from the surrounding rurality, in the gardens and parks, the space was endowed with ornamental plants devoted to the cleaning of and contemplation by a new type of urban dweller.

Plaza de Los Mártires and other small squares

Having transformed the two most iconic squares in Bogotá into parks, the gardening process was extended to other important colonial squares. The ancient *Huerta de Jaime*, one of the oldest places of Bogotá, was also transformed into a garden following the same symbolic elements present in the others. However, as in the case of the Plaza Santander, the symbolic transformation of this square began before any planting occurred. In 1850, the city government decided to change the site's name to the Plaza de Los Mártires on behalf of several heroes executed there during the War of Independence. In order to commemorate those sacrificed lives, they built a 'modest stone column' with the names of those heroes carved on it.[45] Whilst the obelisk had been surrounded by grasses and weeds for 30 years, work to transform it into a garden began in 1881. To achieve this, a French-like garden with geometrical patterns, probably designed by the Italian opera singer Enrique Rossi, was established in the square (see Figure 4.8).

In addition to the city's most important squares, some efforts were made to transform them into gardens and other minor colonial squares in Bogotá. The Plaza Quesada (Quesada's Square), which later received the body remains of the city founder, Gonzalo Jiménez de Quesada, was transformed into a small well-cared garden from 1891 to the first years of the 1920s, when it was demolished as part of the 26th street widening project.[46] Another example was the Jardín del Observatorio (Astronomical Observatory Garden), which after being used by the *Expedición Botánica* team at the beginning of the nineteenth century, was recovered and decorated with bedding plants in 1888, apparently under the direction of Casiano Salcedo.[47] Due to its incorporation into the presidential and Congress house complex in the first decades of the twentieth century, this green space is the only remaining garden among those produced in that period.

In contrast to the systematic gardening undertaken in the central squares of the city, other peripherical squares embellished with flowering plants and trees, such as the La Capuchina and Egipto underwent a rather intermittent and disorganised ornamentation process,[48] Converserly, some larger and more relevant squares did not experience any gardening change.

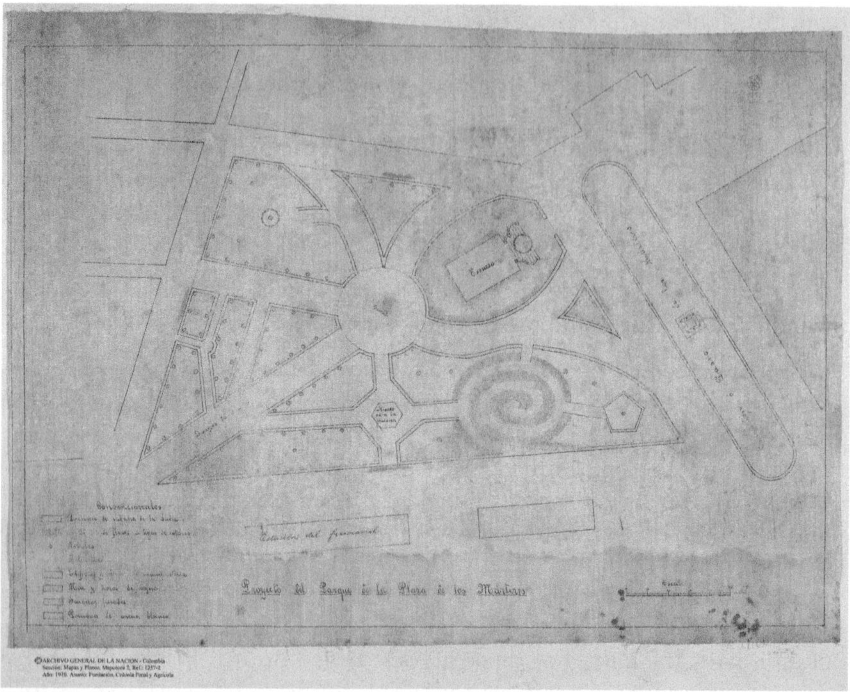

Figure 4.8 Anonymous, *Park layout in the Plaza de Los Mártires*, ca. 1850, watercolour on paper.

Source: Courtesy of the Archivo General de la Nación

The clearest example of the latter is the Plazuela de San Victorino (San Victorino's Square). Located on the west side of the city, this square was historically used as a collection point for merchandise such as raw materials for construction and honey coming from the lowlands.[49] Given its use as a dry port, the establishment of flower beds and trees in San Victorino [as has been commonly named] was never included among urban planning goals.

Some smaller squares that lost their symbolic importance after the colonial period and disappeared with the nineteenth-century compaction of Bogotá include the Plazuela de la Carnicería (Butcher's Square), Plazuela de Guasca[50] or Plazuela de la Aguadita.[51] Their disappearance reveals how the creation of the modern environments in Bogotá was part of a broader process of urban resignification according to modern and Republican values, whereby some places were transformed into gardens and others were dissolved by the city's growth.

El Parque Centenario

Yet, the symbolic and physical changes brought about by the modernisation of Bogotá did not only occur on top of colonial-built environments. Spaces belonging to rural realms at the edge of the city, as well as some residual and transit spaces previously inhabited by spontaneous grasses and weeds, were also transformed into public spaces. Therefore, their spontaneous flora was replaced by an organised, managed and controlled vegetation that better reflected modern perceptions of what the city must look like. A prime example of this transformation is the creation of the Parque Centenario (The Centenary Park), Bogotá's first real public park.

The Parque Centenario was built in a solar covered by weeds and grasses that belonged to the San Diego Franciscan monastery (see Figure 4.9). Historically recognised as Plaza de Marte (Mars Square) and later as Plaza de San Diego (San Diego Square), this site was perceived as a suitable place

Figure 4.9 Edward Walhouse Mark, *Recoleta de San Diego* [San Diego quarter in Bogotá], 1843, watercolour on paper, 17.3 × 25 cm.

Source: Courtesy of the Banco de la República's Art Collection for the construction of a modern park nearly two decades before the commencement of the city's full-scale garden transformations. Initial works undertaken under the supervision of then-Mayor Cenon Padilla included the creation of a one-hectare lake fed by the waters of the San Diego Brook in 1862. Intended as a modern entertainment, the lake was accessorised with a boat capable of carrying 12 people, and *allées* bordered by roses and willows were created along its shores.[52] This early experiment to create a recreational space for the people in Bogotá reveals how the modern necessity for this kind of spaces emerged well before the systematic transformation of the city's squares. However, due to economic constraints, the park in the Plazuela de San Diego was abandoned and some sections were turned into an abandoned and dirty place. Eventually, as so commonly occurs in tropical environments when they are abandoned by humans, spontaneous flora reclaimed the space and weeds and wild aquatic plants replaced the willows and roses.

In 1874, 12 years after the initial effort to transform the space, an ordinance was promulgated seeking to create a park endowed with a fountain and walking paths. However, this project was abandoned, having failed to secure the support of those 'citizens concerned for the city's embellishment', although the Italian architect Mario Lombardi had already received his payment. The idea revived again in 1881, when the Conservative Party took power and prioritised the park's construction as a propagandistic opportunity to strengthen the population's sense of nationalism using the commemoration of the first centenary of Simón Bolívar's birth as a central motif. In contrast to previous attempts to modernise this space, the project's political importance was denoted by the national government creating a committee comprising several recognised elite citizens to coordinate the celebration and the construction of the park. The committee hired the services of another Italian, the architect Pietro Cantini Loi, to produce the blueprints (see Figure 4.10). Thus, whereas the gardens that transformed the ancient colonial squares were produced by the local self-taught gardener Casiano Salcedo, the construction of the Parque Centenario was headed by a Florentine architect. The preference for a European's work over local gardening knowledge reveals the local administration's commitment to endowing the space with strong symbolic prestige. In short, despite the country's endemic economic constraints, the park was completed on time and opened to the public on the 24th of July 1883 as part of a large celebration.[53] Three years later, the park was described as 'the biggest, the most beautiful, the best oxygenized because of its pure airs, and the favourite of the elegant society for their walks' (see Figure 4.11).[54]

Whilst previous investigations have explored the historical details of the construction of the Parque Centenario,[55] my intention here has been to

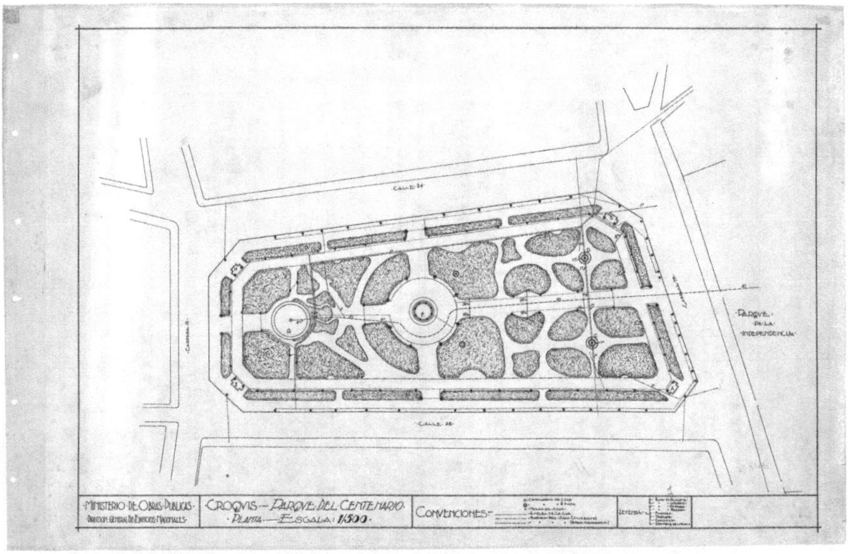

Figure 4.10 Ministerio de Obras Públicas [Ministry of Infrastructure], Parque Centenario layout, ca. 1890, ink on paper.

Source: Courtesy of the Archivo General de la Nación

highlight the space's symbolic relevance. On the one hand, the construction of the Parque Centenario marked the end of the initial stage of the creation of green spaces in Bogotá based on the transformation of colonial squares into gardens. Different to the parks established in former Colonial squares, the construction of the Parque Centenario in the thus far rural environment represents how modern Bogotá advanced over a ruderal and rural-like environments not only through the concrete, bricks and tiles but through an organised nature formed by ornamental plants. In this process, there was a systematic exchange of dandelions and cloves by more urban species such as roses and carnations. The city did not expand only with the urbanisation but expanded its associated flora. The Parque Centenario also reveal the attempts of the city to break its colonial boundaries with its surrounding rurality and how the idea of a modern park not necessarily linked to the colonial square was already deeply rooted as an essential part of the city. Therefore, the city in expansion also represents an amplification of a particular way to perceive, understand, and perform with plants as a reflection of modern urban life and its ways to understand the human body, leisure time, and social control.

Figure 4.11 Henry Duperly, *Parque Centenario*, 1883, albumen print, 18.5 × 23.8 cm.

Source: Dubail, C., and M. Dubai-Acero, eds. *Bourgarel, le Colombien: Voyages d'un diplomate français dans la Colombie du XIXe siècle*. Paris: EdiSens, 2017. Courtesy of the Association Ernest Bourgarel

El Parque de la Independencia

In the transition to the twentieth century, the Latin American city underwent a process of expansion and openness linked to changing forms of understanding and constructing public space. These cultural and spatial transformations would also modify the relationship between plants and people. Bogotá was not indifferent to this process of change, which became evident in 1910. This date marks the centennial of the first attempts at independence and the consolidation of the republic. The celebration of the centenary of independence was an event that was commemorated, with its different nuances, across almost the entire continent.[56] In many cities, this celebration included a new form of green space and included the opening of various parks.

As part of the centennial in Bogotá, the Parque de la Independencia (Independence Park) was opened in 1910 as the venue for the Industrial and Agricultural Exposition. This event tried to emulate the international exhibitions in London in 1851 and Paris in 1877.[57] In this new park, the political connotation that had characterised the gardens in the city did not disappear. The name and the date, which coincided with the centennial of Colombia's independence, are proof of this. In addition, the Parque de la Independencia would be the third green space with the figure of the *Liberator* Simón Bolívar: this time, in the form of an equestrian statue. Despite the statue of the Liberator and in spite of its name, The Parque de la Independencia would be widely remembered as the space designed for the celebration of modernity that brought with it capitalist ways of life. In the 1910 exhibition, people from Bogotá witnessed the modern innovations of industry that were changing daily life through, for example, the mass production of quotidian elements such as soaps or candles thus far produced locally with vegetal elements from the forest.[58] This exhibition took place in the park and included the construction of several pavilions, including the Light Pavilion, which, for the first time, demonstrated the existence of electric light inside enclosed spaces.

Unlike the garden in the plaza or the Parque Centenario, the Parque de la Independencia was an open place of large dimensions. It was not enclosed by forged fences brought from Europe. Its careful landscape design, carried out by the Japanese gardener Tomohiro Kawaguchi, made the visitor feel immersed in a space that, while not completely rural, resembled the landscape.[59] Consequently, In the Parque de la Independencia, people no longer orbited around the patio/garden or the square; there, they were no longer limited. It is in this park that the citizens found themselves immersed in a space created with plants for the first time; there, the individual was the very centre of the experience of the place. In this public park, there were blind spaces that simulated the bucolic atmosphere of rural nature, now lost. This park responded to but also promoted new ways of being in a space created with plants.

A remarkable characteristic of the Parque de la Independencia is that it represents the beginning of a relationship between activities of the body and green spaces. After the end of the exhibition, a skating rink was built in the Industrial Pavilion, while the Egyptian Pavilion hosted a hall for physical education. In the city's first gardens built in the squares, the hygienic power of plants worked on the moral and physical reconstitution of the bodies. In contrast, in a big open space such as the Parque de la Independencia, the body of the now factory workers found, through activity and exercise, a way to produce health. The body regulates itself, and unlike gardens in patios and squares, plants did not necessarily generate a healthy space. In the Parque de la Independencia, the physical and moral health emanates from the inside. It is an internal force, just like the invisible force that made the bulbs glow in the Light Pavilion.

This new way of relating to green spaces strongly bound to the human body found a culmination through the practice of sports. The body, transformed into the centre of the world, is exercised and maintained, just as had been the case for the soul in the colonial city. The new ways of in which bodies relate to the city and, specifically with open spaces would also modify the relationship of citizens with plants, especially with grasses.[60] In the colonial and republican cities, grasses had been exclusively for animal consumption in the city's fields and for human consumption through the cultivation of wheat, barley, and oats. But in the city of the late nineteenth and early twentieth centuries, grasses, usually introduced from Africa, were used as a field, as a playground. Plants were no longer only edible or ornamental; they became vegetal carpets where the human body was exposed to injuries in sports that simulated the competition of the capitalist society already established in Latin American cities. This relationship crystalised in the creation of stadiums, which paradoxically turned out to be closed spaces open again to the sky, just as the patios. There, people enjoyed the spectacle of sport sitting around a green space, namely a pitch.

In sum, by the begining of the twentieth century Bogotá had been endowed with several gardens built upon the colonial squares and two relatively bigger parks established in the outskirts of the city. The creation of gardens in Bogotá reflects as any other the creation of the public life linked to the capitalist ways of living. Closing the nineteenth century, the city as a spectacle was clearly defined, thus transforming the traditionally built relationships between people and plants.[61] However, the city stage demanded the intervention of people able to craft space through the manipulation of plants. The next section explores the people who directly or indirectly contributed to the creation of this photosynthetic scenery.

Notes

1 Tomas Carrasquilla, "Enredos e Incongruencias," in *La Ciudad y Sus Cronistas* (Medellín: Instituto Tecnológico Metropolitano, 2009), 245.
2 Correspondence between Genaro Valderrama and Ministerio de Hacienda, 4 Abril 1897, folio 114, tomo 823, fondo del Ministerio de Obras Públicas, AGN.
3 Rio de Janeiro, for example, had several large urban transformations: overwhelmed by Napoleon I, King Juan VI of Portugal relocated his royal residence to Rio do Janeiro. In order to bring royal splendour to the Brazilian city, French urban planners such as Auguste Grandjean de Montigny and Auguste Glaziou opened new streets and refurbished others, as well as creating new squares and parks, see Francis Korn and Lidia de la Torre, "La Vivienda En Buenos Aires 1887–1914," *Desarrollo Económico* 25, no. 98 (Julio 1, 1985): 245, https://doi.org/10.2307/3466807; Romero, *Latinoamérica las ciudades y las ideas*, 223.
4 Mejía, *Los Años Del Cambio*, 38.
5 Julio Vergara y Vergara, *El Desarrollo Urbano de La Capital. Las Obras Del Cuarto Centenario* (Bogotá: Editorial Nueva, 1936), 6.

6 Estela de la Merced Restrepo Zea, *El Hospital San Juan de Dios 1635–1895: Una Historia de La Enfermedad, Pobreza y Muerte En Bogotá* (Bogotá: Universidad Nacional de Colombia. Facultad de Ciencias Humanas. Departamento de Historia. Centro de Estudios Sociales (CES), 2011).

7 Although Bogotá experienced an overall population increase over the course of the nineteenth century, German Mejia Pavony has shown how growth occurred through two different pulses, one between 1801 and 1848 and a sharper increase between 1870 and 1912. Mejía, *Los Años Del Cambio*, 227–55.

8 Jorge Brisson, *Viajes Por Colombia En Los Años 1891 a 1897* (Bogotá: Imprenta Nacional, 1899), 158.

9 Mejía, *Los Años Del Cambio*, 139–41.

10 Restrepo Zea, *El Hospital San Juan de Dios 1635–1895*, 260.

11 Steuart, *Bogotá in 1836–7: Being a Narrative of an Expedition to the Capital of the New-Granada and a Residence There of Eleven Months*, 134.

12 By 1883, all the capitalist production was represented by: six factories, eight banks, one insurance company, one theatre, 21 hotels, 30 official newspapers, one company of gas-based public illumination, one factory of glass, 20 presses, ten binding shops, 60 ateliers and shoe, 50 tailoring, 40 saddleries, 12 watch shops, 15 jewellery, 25 smithy, 20 sheet metal shops, 100 carpentries, 20 carpet ateliers, 16 hairdressing shops, 15 dentist, 30 pharmacies, 12 dressmaking shops, four photography shops, two lithographs, ten carving offices, two of sculpture, 24 of painting, some architects, and endless masonries; 35 commercial agencies, 16 carriage agencies, 12 fretting agencies, two musical agencies, four mortuary agencies, four auction agencies, more than 150 solicitors, 50 doctors and 20 engineers, see Pérez, *Geografía General de Los Estados Unidos de Colombia y Geografía Particular de La Ciudad de Bogotá*, 402.

13 Miguel Samper, *La Miseria En Bogotá* (Bogotá: Universidad Nacional, [1898] 1969).

14 A good example of the interaction between morality and health is expressed in the opinion of the French physician Thomas Dagoumer (1762–1833/35) who invoked poverty, ignorance, and immorality to explain the high mortality rate among the *ille de cité* in Paris, see Nicholas Green, *The Spectacle of Nature: Landscape and Bourgeois Culture in Nineteenth Century France/Nicholas Green* (Manchester: Manchester University Press, 1990), 50–51.

15 Jon A. Peterson, "The Impact of Sanitary Reform upon American Urban Planning, 1840–1890," *Journal of Social History* 13, no. 1 (1979): 83–103; Luis González, *Medellín, Los Orígenes y la Transición a la Modernidad: Crecimiento y Modelos Urbanos 1775–1932* (Medellín: Univ. Nacional de Colombia, 2007).

16 P. Ibañez, "Sesión del 14 de junio de 1886," *Revista Médica*, July 20, 1886.

17 Drawing on Foucault's ideas on the pulmonary nature of nineteenth century humans, plants were broadly accepted as organic filters capable of contributing to the health of cities, see for instance Michel Foucault, *The Birth of the Clinic: An Archaeology of Medical Perception* (New York: Vintage Books, 1975); Karen R. Jones, " 'The Lungs of the City': Green Space, Public Health and Bodily Metaphor in the Landscape of Urban Park History," *Environment and History* 24, no. 1 (February 1, 2018): 39–58, https://doi.org/10.3197/096734018X15137949591837; Matthew Gandy, *Concrete and Clay : Reworking Nature in New York City/Matthew Gandy, Urban and Industrial Environments* (Cambridge, MA: MIT Press, 2002); David Harvey, *Paris, Capital of Modernity* (New York and London: Routledge, 2003); Henry W. Lawrence, *City Trees:*

A Historical Geography from the Renaissance Through the Nineteenth Century (Charlottesville and London: University of Virginia Press, 2008); Richard Sennett, *Flesh and Stone: The Body and the City in Western Civilization* (London: Faber and Faber, 1994); Anna Wilson, "Uprooting Melbourne. A Story of a City as Revealed by Trees" (PhD diss, University of Tasmania, 2014).

18 Francisco Arango and Francisco Uribe, "Cementerios: informe de una comisión," *Anales de la Academia de Medicina de Medellín*, June 1889.

19 Francisco Uribe, "Higiene local," *Anales de la Academia de Medicina de Medellín*, February 1888.

20 Correspondence between Sebastian Castell and the Minister of Economic Affairs, 22 Mayo 1889, folio 56, tomo 828, fondo Ministerio de Obras Públicas, AGN.

21 "Bibliografía. Higiene Pública y Privada," *Revista Médica*, June 20, 1886.

22 Francisco Javier Santero, *Elementos de Higiene Privada y Pública*, vol. II (Madrid: El Cosmos Editorial, 1885), 13–14.

23 For Bogotá, however, the discussion around the best kind of trees to be used on the street, with very few exceptions, was a topic not addressed by doctors. The discussion about the best trees to be planted in Bogotá came during the first decades of the twentieth century as a response to the conflict that emerged after the adoption of the eucalyptus as the typical urban tree.

24 About the creation of the public as a consequence of the modern and capitalist way of life, see Marshall Berman, *All That Is Solid Melts into Air. The Experience of Modernity* (London: Verso, 1982).

25 Claudia Cendales in her detailed history on Bogotá's parks has suggested that the creation of gardens in t in Bogotá's squares was more influenced by London than Paris. Since the seventeenth century some squares in London were adorned with plants, endowed with urban facilities and enclosed by iron fences Cendales Paredes et al., *La Vida Privada de Los Parques y Jardines Públicos*; see also Henry W. Lawrence, *City Trees: A Historical Geography from the Renaissance Through the Nineteenth Century* (Charlottesville and London: University of Virginia Press, 2008).

26 Claudia Cendales, "Un parque extenso y amplio para dotar con él a nuestra querida capital': La exigencia de la creación de un parque y el panorama del arte paisajístico a finales del siglo XIX en Bogotá," *Paisagem e Ambiente*, no. 29 (October 8, 2011): 27, https://doi.org/10.11606/issn.2359-5361.v0i29p25-38.

27 Julio H. Bonilla Romero, "Aproximaciones al observatorio solar de Bacatá-Bogotá-Colombia," *Revista de Topografía AZIMUT* 3 (2011): 9–15; Julio H. Bonilla Romero, Edier H. Bustos Velazco, and Jaime Duvan Reyes, "Arqueoastronomía, alineaciones solares de solsticios y equinoccios en Bogotá-Bacatá," *Revista Cientifica* 27 (January 2017): 146–55.

28 What Lefebvre would have called the city's main *lived space* Henri Lefebvre, *The Production of Space*, 1st ed. (Oxford, UK: Wiley-Blackwell, 1991). Nowadays the square is bordered by the Palace of Justice to the north and the National Capitol to the south, and the Primary Cathedral of Bogotá and the Liévano Palace, the seat of the city's mayor, are situated on the eastern and western sides, respectively.

29 José Ortega y Gasset, *El tema de nuestro tiempo; la rebelión de las masas* (México: Porrúa, [1938] 2005), 195–96.

30 Fabio Zambrano, "La Ciudad Como Representación Política," in *Publicistas, Prensa y Publicidad En La Independencia de Hispanoamerica*, ed. Moisés Guzmán (Morelia: Universidad Michoacana de San Nicolás de Hidalgo, 2011), 177.

31 Miguel Cané, *En Viaje* (Caracas: Fundación Biblioteca Ayacucho, 2005), 191.
32 André, "La Vuelta Al Mundo, Nuevo Diario de Viajes. La América Equinoccial' (Colombia-Ecuador-Perú)," 242.
33 Mejía, *Los Años Del Cambio*, 185.
34 Ibid.
35 Delgadillo, *El Parque del Centenario en Bogotá: Transformación urbana, itinerario y significado*, 42.
36 "El jardín de Bolívar," *El Conservador (Bogotá)*, Agosto 10, 1882.
37 "El Jardín de Bolívar," *El Conservador*, Septiembre 16, 1882.
38 Rubén Hernández Molina and Fernando Carrera Zaldúa. *Las Nieves, La cuidad al otro lado* (Bogotá: Instituto Distrital de Patrimonio Cultural, 2010), 13–14.
39 Daniel Ortega, *Historia Del Parque Santander* (Bogotá: Sociedad de Mejoras y Ornato, 1926).
40 Lisímaco Palau, *Guía Histórica y Descriptiva de La Ciudad de Bogotá* (Bogotá: Imprenta de Vapor de Zalamea, 1894).
41 Ibid., 37.
42 In 1881, Cané also mentioned how the Plaza de Bolívar was 'a square of one block size, having no trees or bench, cold and deserted'. This description confirms that the first place to host trees and other plants in Bogotá was the Parque Santander, contrary to what has been assertions by researchers such as Claudia Cendales that the first park in Bogotá was built in the Bolívar Square, see Miguel Cané, *En Viaje* (Caracas: Fundación Biblioteca Ayacucho, [1883] 2005), 185; Cendales, "Un parque extenso y amplio para dotar con él a nuestra querida capital."
43 Ortega, *Historia Del Parque Santander*, 38.
44 André, "La Vuelta al Mundo, Nuevo Diario de Viajes. La América Equinoccial (Colombia-Ecuador-Peru)," 240.
45 Ignacio Borda, *Monumentos Patrióticos de Bogotá. Su Historia y Su Descripción* (Bogotá: Imprenta de la Luz, 1892), 32.
46 Delgadillo, *El Parque del Centenario en Bogotá: transformación urbana, itinerario y significado*, 36.
47 Ibid., 40.
48 Ibid., 34–39.
49 Mejía, *Los Años Del Cambio*, 187.
50 While *guasca* in Colombia is a name for the plant *Galinsoga parviflora*. However, according to the Chibcha language spoken by Muiscas, it also means "skirt of the mountain range," see http://muysca.cubun.org/Gram%C3%A1tica_de_Lugo/fol_7r.
51 Mejía, *Los años del cambio*, 189.
52 Cenón Padilla, "Diarios de trabajos de la alcaldía del cuartel de las nieves," *El Colombiano (Bogotá)*, Agosto 7, 1863.
53 Delgadillo, *El Parque Del Centenario En Bogotá: Transformación Urbana, Itinerario y Significado*.
54 Correspondence between Eugenio Moreno and the Minister of Economic Affairs, 22 May 1886, folio 67, tomo 823, fondo Ministerio de Obras Públicas, AGN.
55 Claudia Cendales, "Los Parques de Bogotá: 1886–1938," *Revista de Santander*, 2009; Cendales, "Un parque extenso y amplio para dotar con él a nuestra querida capital"; Hugo Delgadillo, *El Parque Del Centenario En Bogotá: Transformación Urbana, Itinerario y Significado* (Bogotá: Alcaldía Mayor de Bogotá. Secretaría General. Dirección Distrital de Archivo de Bogotá, 2019); Maria Guerrero, "Pintando de verde a Bogotá: visiones de la naturaleza a través de los

parques del Centenario y de la Independencia, 1880–1920," *Revista de Historia Ambiental Latinoamericana y Caribeña* 1, no. 2 (2012): 112–39; Mejía, *Los Años Del Cambio*.

56 For an example of the centenary celebration see Mauricio Tenorio, *I Speak of the City* (Chicago: University of Chicago, 2012).

57 Cendales Paredes et al., *La Vida Privada de Los Parques y Jardines Públicos*, 102.

58 Ibid., 122.

59 Ibid., 111.

60 José Luis Pardo, *Las Formas de La Exterioridad* (Valencia, España: Pretextos, 1992).

61 About the city as a spectacle, see Green, *The Spectacle of Nature*; David Harvey, *Paris, Capital of Modernity* (New York and London: Routledge, 2003).

5 The gardeners

In contrast with the flora of the colonial and early republican city which was either cultivated in domestic spaces or freely thriving in open spaces and residual areas, the new gardens and parks demanded some formal gardening skill. As a primary response to the lack of gardening and horticulture tradition, the government sought solutions from foreigners that, with the integration of Colombia into the international trade system, started to appear more often in Bogotá. From the mid-nineteenth century a vast range of people including fortune seekers, Humboldt's emulators, businessmen, engineers, professors, plant hunters and some professional gardeners visited the city.[1] Although limited in comparison with the work of local and self-taught gardeners, some of these foreigners contributed in various ways to the foundation and development of Bogotá's modern green spaces. These foreigners embodied particular conceptions about urban nature that were translocated into entirely new contexts. Therefore, comparable to the European city planners who changed the face of Latin American cities, these figures represent the gardening side of 'transnational and mobile urbanism'.[2] Carrying specific ideas about the city, these trained and improvised gardeners embodied a very specific way to conceptualise the role of plants as part of the modern city. This section presents some of the most relevant characters that contribute in one way or another to the consolidation of the gardening practices in Bogotá.

The Italians: Enrique Rossi-Guerra and José Fabi

Not all of the foreigners who contributed to the modern construction of green spaces in Bogotá received formal education in horticulture or gardening. The most outstanding example is the improvised role of the Italian opera singer Enrique Rossi-Guerra as a garden designer for public spaces. As part of the aperture into the world and the spectacles produced by modern capitalist society,[3] the *Banzzanni* Lyric Company came to Bogotá from Italy in 1858. This was the first opera group to arrive in the country. Among

DOI: 10.4324/9781003208167-8

its performers was the tenor Rossi-Guerra, who had been well-recognised since 1830 for his performances in Mexico City, Chile, Argentina, and Peru. Rossi's prestige in Colombia grew considerably when he conducted one of the most acclaimed opera companies, the Rossi-d'Achiardi. Following his success in the country, he made Bogotá his permanent residence in 1872 and opened an Italian restaurant. However, when the Conservative Party headed by Rafael Nuñes took over in 1880, one of its polemic decisions was the flattening of the *Maldonado* theatre, which had historically been devoted to scenic performances. As a consequence, Rossi's presence as an opera tenor on Bogotá's public scene was severely diminished. However, Rossi ultimately made a career for himself as a gardener and landscape designer.[4]

Following the success of international agricultural exhibitions in Cologne (1865) and London (1879), national exhibitions become a traditional (and repetitive) event in Colombia. In accordance with this trend, Bogotá hosted its Second National Agricultural Exposition in 1881. Products from distant regions of the country were displayed, and various types of natural resources ranging from tobacco to rocks, as well as curiosities such as archaeological treasures, were transported for days on beasts of burden for presentation at the capital. These exhibitions served as a means to obtain an inventory of products in regions that remained only weakly connected to the international market, and like their European counterparts, their aim was primarily business-related. Consequently, they attracted the attention of a considerable part of the Bogotanian sector that was interested in Colombia's integration into world commerce *via raw materials*.[5]

The venue of the 1881 exhibition was a house previously acquired by the government for use as the headquarters for the College of Agriculture. Although the press declared that particular exhibition a failure due to the low quality of the exhibited products, its organisers nonetheless gained great experience. Notably, in association with this event, the name Rossi-Guerra once again appeared in the city's newspapers; this time not as a famous tenor, but rather as a member of the exhibition's organising crew who had been entrusted with the task of 'embellish the place with the gardens and take care of all objects'.[6]

Later, Rossi's name appeared again as part of the campaign to transform the old colonial squares into modern park spaces. The Italian had been hired by the government to refurbish the so-called *Huerta de Jaime* and transform it into a garden. Although there is limited historical evidence concerning the role played by Rossi in this urban transformation, the available information shows that as a poor and forgotten old man whose glorious days as one of the most important characters in the city had long passed, Enrique Rossi-Guerra found a place a gardener in the Quinta Segovia (See Figure 2.2). In this country house, he was charged with the care of an old dromedary brought from Algeria by a man named Pedro Navas Asuero. Thus, as portrayed by José Perdomo, these two foreigners 'linked by fatality' found a place to die in one of the earliest established gardens in Bogotá.[7]

There is some irony to be found in Rossi's final days in Bogotá; a person who had spent his most productive days onstage ended up caring for the gardens that had emerged as one of the stage-like manifestations adopted by the modern city. However, beyond any specific life story, what must be highlighted is the lack of trained people who could create and take care of Bogotá's earliest gardens, and how the absence of qualified personnel made any person related to the arts a suitable person to undertake gardening works.

Whereas Rossi-Guerra was part of the initial stages in the greening process of Bogotá, his compatriot, the professional gardener José Fabi, can be considered the last of the foreigners involved in the creation of the city's gardens. Although the figure of Fabi first appeared in Bogotá as late as 1934, a period that exceeds the temporal limits of this research, his role in the city's garden production requires at least a brief review. Fabi's role in the progress of gardening works is presented to highlight how the external presence in the construction of Bogotá's green spaces stretched into the latter decades of the first half of the twentieth century, thus demonstrating the city's continued underdevelopment in this area.

José Fabi was hired at the beginning of the 1930s by the Gardens and Parks office belonging to the Ministry of Infrastructure. Known either as the Technical City Gardener or as the National Park Agronomist, he was responsible for the most sophisticated and high-level gardening practices the city had implemented thus far. During Fabi's tenure many unattended and residual urban spaces were used in the construction of gardens. For instance, whereas flower beds and small bushes had become part of Bogotá's road system, the flourishing of small gardens scattered around the city reveals high levels of care (see Figure 5.1). The strengthening of the city's gardening practices was possible thanks to a group of knowledgeable gardeners under the Italian's supervision.

The city's gardeners gained a new pride in their work due to the knowledge they gained under Fabi's supervision, and the staff's emerging self-identification as trained gardeners led to the creation of Bogotá's Gardeners Union, the first organisation of this nature in Bogotá since the transformation of the colonial squares into green spaces in 1880. This local body was not only a deliberative power in charge of advocating for their labour conditions but also a defender of a particular gardening aesthetic.[8] Although the political will responsible for the strength of gardening practices manifested in Fabi's works is not presented here, its consequences created a sort of fashion for the production of gardens, and both public and private institutions showed an unusual interest in gardens in the 1930s.

Figure 5.1 Anonymous, Plazuela Las Nieves, ca. 1933, postcard.
Source: Courtesy of the Biblioteca Pública Piloto de Medellín

The Ministry of Infrastructure and particularly the Gardens and Parks office became the centre of gardening knowledge in the country, and public institutions such as the Universidad Nacional and other public institutions sent letters to the ministry asking for gardeners to accomplish works there.[9] Such works added pressure to gardening and park administration, which meant a greater burden for gardeners who now had to do more with the same resources. Gardeners, thus, manifested their discomfort with their work overload through the gardeners union on several occasions. However, the success of Bogotá's gardening practices under Fabi had not attracted the attention of only public institutions: in 1934, the Ministry received a petition from the prestigious hotel Apulo in the hot weather city of Girardot requesting his presence to create a garden there.[10]

Fabi's critical role as the head of the gardening guild is perhaps best unveiled by the response of the city's gardening staff when he had to undertake a trip to Rome to visit his mother in 1936. On this occasion, the entire staff of city gardeners, which comprised around 35 men and six women, sent a letter to the minister asking to keep Fabi's position during his absence and try to persuade the Italian to return:

> The government will be deprived of the knowledges of the outstanding professional and tireless administrator whose work has brought many

> benefits to our country and, particularly to the labours that having had the fortune to work under his direction we will not forget the constant interest presented by Mr Fabi looking for improving us in our profession. . . . It is quite true that our guild has acquired valuable knowledge and perfection in the gardening and agriculture practices in general due to the presence of Mr Fabi. It is also true that many of us have upgraded from simple gardening peons to the most skilled agricultures mainly because of the education received from our benefactor.[11]

Interestingly, the Italian contribution to the gardening of Bogotá represents two tips of history. Whilst the amateur work of Enrique Rossi-Guerra in the decade of 1880s was a reflection of the improvisation and the lack of highly skilled gardeners, José Fabi's works show how the twentieth century saw the professionalisation of gardening practices, which in turn reflects the consolidation of a particular urban flora in Bogotá different to its surroundings. However, between these two extremes embodied by Italian nationals, many others contribute to the development of gardening in Bogotá.

The British: Robert Thomson

In 1893, a British named Robert Thomson arrived in Bogotá. Thomson had completed horticultural training at the Kew Royal Botanical Garden in 1862, which qualified him to serve as an assistant gardener at the newly established Castleton Botanical Garden in Jamaica until 1867. Although it seems that Thomson visited Colombia several times after his stay in Jamaica, his presence in Bogotá in 1893 was the most significant in the context of this discussion. Although Claudia Cendales [one of the few researchers addressing the gardening history in Colombia] has suggested Thomson was invited to the city by the Ministerio de Fomento Development Minister, the letters on which this claim is based do not provide sufficient proof to affirm that his presence in the country was due to institutional will.[12] It seems that Thomson was used more as a gardening consultant than as a direct employee of Bogotá's municipality.

Thomson's ephemeral presence in Bogotá left a rich testimony to the city's green spaces. In a series of letters with the Ministerio de Fomento (Ministry of Development), Thomson presented himself as a voice of authority, and his opinions were considered the first evaluations of the city's parks and public gardens through the eyes of a European 'expert'. Thomson's diagnosis of the city's parks and gardens was highly critical and was presented in a 13-page report-like letter [first written in English and then translated to Spanish] sent to the Ministry on the 30th of January 1893. Thomson focused his diagnosis on two related topics: i.) the parks' design and ii.) the plants used in the city ornamentation.

Regarding the former, Thomson expressed complete disagreement with the general design of most of the city's parks, including the inclusion of pathways:

> In all the public parks, named that on the Plaza Bolívar, San Francisco, Centenario and Los Mártires a great mistake has been committed in their formation, I refer to the walks which immediately bound on this iron-railings. Thus, instead of these walks immediately within the fences, shrubberies. etc., ought to have been established- Not a walk skirting a street. From a gardening point of view, these spaces occupied as a walk are not only objectionable, but, also, a large amount of wasted spaces is the result. Sometimes as much as from 16 to 20 feet in width all around the fence.[13]

In his comprehensive diagnosis of each of the green spaces in Bogotá, he first highlighted their main failures and then proposed a series of corrective measures. Addressing the Parque de Los Mártires, for instance, Thomson mentioned how: 'Like the other parks the chief defect in the formation of this is the great broad walk bounding the railing', and he suggested 'The broad walk contiguous to the railing should be converted into ground for planting purpose. And several of the other walks should be made narrowed'.[14]

In a further letter sent by Thomson to the Minister A. Torres in September, he specified 'the works I've planned to be necessary for the re-formation of these parks'. Notably, Thomson proposed the use of plant borders that was then very popular in British gardens as a solution to address the issue of the paths in the Parque Centenario. In embodying the English way of creating gardens, Thomson proposed that:

> All the walks in this park, with the exception of the one skirting the fence, have been recently put in good order. The work, which I am prepared to undertake would not in any way interfere with the renewed walks. One of the most important improvements to be made in this park is to convert the walk skirting the fence into a border for plant cultivation. In my letter. . . . I fully explained the inutility of a walk bounding a fence—a defect that applies to all the parks here. Nearly all round this park the walk is about 20 feet wide, hence a very large proportion of the area of the park will be reclaimed from a useless waste into a plant border.[15]

Thomson's recommendations were not only aimed at parks and public gardens. He also suggested ways in which the city's promenades and tree-planted roads could efficiently correspond to the design of such places in Europe. Regarding the Camellón de las Nieves (Las Nieves Road) he explained that 'Symmetrical planting must be the leading feature of avenues. This has not been pursued in Bogotá as the most diverse and disproportionate forms of plants have been set'. This comment reveals how the organisation and control of nature synonymous with gardening practices in Europe was at that time hardly implemented in a city of residual spaces generally endowed with spontaneous flora.

Of course, the lack of alignment between what should be and what was established as gardens in Bogotá as seen through Thomson's eyes included the city's plants. Although the next chapter will provide a more detailed examination of Thomson's opinions concerning the most suitable plant species, it is important to mention that Thomson perceived plants as an integral component of the layout of parks and avenues. For instance, he was a severe critic of the eucalyptus trees broadly used in the decoration of parks (see Figure 10.2). When discussing the Parque Santander, he suggested that 'certainly all the *Eucalyptus globulus* should be cut down, but before doing it a large stock of choice plants should be prepared [*illegible*] as to plant systematically and renovate the park'. Equally, concerning the Parque Centenario, he expressed:

> The row of eucalyptus both outside and inside the railing at present form the most striking feature of the park. A few more years would render them unsightly (*See Figure 10.2*). The Eucalyptus groves to a gigantic size and its over-spreading branches, together with their wider extended roots are prejudicial to nearly all kind of plants adjacent to it . . . with the exception of the eucalyptus and about 100 pines the majority of the latter comprising only two or three, all the other plants have but little effect and value.

In order to set up adequate green spaces in the city, Thomson proposed ambitious plans with an overall cost of 13,000 Colombian pesos [pesos hereinafter].[16] For example, to transform the San Diego walk by the fence into a plant border, he asserted that between 'two and three thousands of good soil at a cost of, probably, three *reales* each would be necessary'.[17] Bringing what he had seen at Kew, Thomson also suggested the construction of a greenhouse artificially heated in the same park, which would be used for propagation purposes as well as a source of recreation and instruction. Discussing the greenhouse, which would cost 3,000 *pesos*, Thomson also noted that 'When the parks are planted and completed [according to his proposals] the expense of keeping them in good order would be for annum 2,400 *pesos*'.[18]

Thomson's proposed greenhouse and annual park maintenance costs reflected the lack of correspondence between his gardening knowledge acquired in one of the most powerful institutions of the British Empire and the economic context of a country like Colombia. A 3,000 *pesos* greenhouse and a 2,400 *pesos* annual maintenance expense were untenable for a city that did not expend more than 5,000 *pesos* for the yearly maintenance of all of its parks combined. It is perhaps as a consequence of this lag that apart from his detailed diagnostic and suggestions were ultimately rather marginal, and his ideas to convert them into proper English gardens were never implemented. The only verifiable register about the work made by Thomson was a couple of jalapa trees [undetermined species] planted by him in the Parque Centenario in 1891 and two Zea trees [undetermined species] in the Parque Bolívar.[19]

Finally, despite Thomson's enthusiasm for developing Bogotá's 'green spaces', his attention was also devoted to other, more profitable concerns. In the same letters sent to the Ministry of Development addressing the situation of Bogotá's parks, he proposed a business involving the importation and acclimatisation of economically useful plants. Thomson proposed to use some hectares of land near the hot weather city of Girardot to establish a plantation where he would engage in the propagation of profitable plants to be used as part of an importation substitution scheme. According to Thomson, these plantations would generate major savings for the Colombian government, which was then expending 'exorbitant' sums of money on the importation of foreign plant products. The list of more than 15 plants included in his project encompassed cardamom (*Elettaria cardamomum*), nutmeg (*Myristica fragrans*), cumin (*Cuminum cyminum*), tea (*Camellia sinensis*), black pepper (*Piper nigrum*), kola nut (*Cola acuminata*) and cloves (*Syzygium aromaticum*), as well as, incredibly, the native rubber trees of Ceará and Pará (*Manihot carthaginensis subsp. glaziovii and Hevea brasiliensis*). Thomson's business proposals highlight the direct relationship between tropical plant diversity and the creation of wealth, which was at the core of the British imperial consolidation throughout the nineteenth century.[20] As in the case of the garden proposals, Thomson's ideas were a vivid representation of a very specific way to understand nature. Finally, despite his suggestions seeking to create real European gardens atop the Andes Mountains, Thomson's place in the history of Bogotá can be considered that of a well-educated fortune hunter. When the figure of Thomson reappeared in 1899, he was not trying to recreate European gardens in Bogotá, but rather had established one of the Colombian Amazon region's foremost rubber plantations with more than 60,000 trees. As revealed by the British human rights activist Roger Casement in 1911, these plantations were part of one of the most shocking cases of modern slavery and genocide documented as part of the massive exploitation of rubber in the Colombian and Peruvian Amazon jungles, which was accomplished throughout the forced labour of enslaved indigenous people.[21]

The Japanese: Tomohiro Kawaguchi

In 1908, wealthy businessman Antonio Izquierdo de la Torre undertook a long journey that took him to Japan and other countries in Asia. As he presented it to the Colombian Parliament, 'I am working on an issue about Japanese migration towards Colombia, especially to bring labours to work in the Pacific coast . . . the workers will be used to accomplish agricultural jobs and in the railway construction'.[22] Izquierdo's journey and labour importation efforts have been interpreted by Toraji Irie as an attempt to counter the influence of the United States in the country following Panamá's independence. According to Irie, one of the aims of Izquierdo's visit to Japan was to request 10,000 workers to be placed in the newly created Panamá border in the north; however, as Ines Sanmiguel has pointed out, there is insufficient primary information to confirm this claim.[23] Regardless of Izquierdo's intentions, the result of his commitment was very discrete: rather than a herd of Japanese people coming to occupy the shores of the Pacific Ocean or living in the wild Panamá border, the overall Japanese migration was comprised of only a single person. Accompanying Izquierdo upon his return to Colombia was the gardener Tomohiro Kawaguchi, who, according to Izquierdo, 'was given by the Count of Okuma with whom [Kawaguchi] had worked four years after working eight years in the imperial gardens'.[24]

It seems that Kawaguchi's presence in Colombia was the result of an assignment given by President Rafael Reyes who needed a gardener to care for the plants on his farm in the town of Chapinero;[25] however, his works were not limited to the president's private property. Kawaguchi's knowledge was useful in the ornamentation of the San Diego Forest, and he appears to have played an important role in the area's landscaping during its transformation into the Parque Centenario.[26] However, his precise role remains unclear, nor is it known whether or not he was ever in contact with the local gardener Casiano Salcedo. In addition to the aforementioned works, the only documentation regarding this Japanese garden concerns his death in a small town in the mountains near Bogotá.[27]

Kawaguchi was not the only Japanese involved in gardening development in Latin America. According to Sergio Hernandez, Tatsugoro Matsumoto was a significant figure in the constitution of urban nature in Mexico City. Having arrived in Mexico in 1896 after creating several Japanese-style gardens for upper-class people in Perú, Matsumoto obtained considerable recognition as a gardener and 'landscape designer'. The dictator Porfirio Díaz commissioned his services for the planting and ornamentation of the Chapultepec Forest in 1910, and after the overthrow of the Díaz regime by the revolutionary government in 1911, Álvaro Obregon hired Matsumoto to oversee the ornamentation of the city's road system. Matsumoto lined the city's roads with thousands of the flamboyant *Jacaranda mimosiflora*, an Argentinian-native tree which brought about a colourful spectacle. This artistic innovation would soon become one of the most important natural symbols of Mexico's capital.[28]

The local and self-taught gardeners: Salcedo and Valderrama

As demonstrated earlier, with the exception of the José Fabi's works undertaken during the 1930s, the role of foreigners in the construction of Bogotá's green spaces was not significant. Although they certainly influenced the ways in which gardens and parks were designed, local and self-taught gardeners assumed the main responsibility for the creation of these spaces. Toward the end of the nineteenth century, two figures emerged as personifications of local gardening skills in Bogotá, they were Casiano Salcedo and Genaro Valderrama.

Casiano Salcedo

Casiano Salcedo was undoubtedly the central figure in the creation and consolidation of Bogotá's gardening development throughout the last decades of the nineteenth century. Salcedo did not have any formal gardening education. Conversely, this self-taught gardener self-recognised as the 'First Scientific Gardeners' of Bogotá, acquired most of his knowledge from books, that he read after learning French because of his 'love to botany'.[29] It is highly likely that Casiano Salcedo had access to some contemporary volumes that are still available in the Biblioteca Nacional; their presence in the country's main library reflects that some gardening books reached Bogotá despite the city's intellectual isolation.[30]

The historical background of Salcedo is obscure; however, he managed to create a solid network that put him in contact with some international nurseries such as the Vilmorin Nurseries in Paris. Additionally, he developed contact with Louis Benoît Van Houtte in Ghent (Belgium) from which he apparently managed to introduce several species of plants such as camelias (*Camellia japonica*) and azaleas (*Rhododendron simsii*) to Bogotá.[31] Additionally, his social network included different personalities from the political life of the country. This is the case of ministers and even President Manuel Murillo Toro, from which Casiano Salcedo obtained seeds of *Eucalyptus globulus* that were profusely planted in Bogotá.[32] Salcedo's international network allowed him to partially materialise what he had learnt and seen in the books. An example of this is the use of Araucarias (*Araucaria heterophylla*) in the Salcedo's ornamentation of the Parque de Bolívar. During the last part of the nineteenth-century Araucarias were popular urban ornamental and as part of cities was widely advertised in gardening and horticulture books. Because of his contacts, Salcedo obtained araucarias seeds and by planting them in the city's park, he managed to crystalise in Bogotá the gardening trend of his epoch.[33]

Although Salcedo cannot be considered to have been part of Colombia's elite, his close contact with them places him in an authorised position as

Bogotá's first and foremost gardener. Consequently, and spite of the presence of trained gardeners such as Robert Thomson, the self-taught Casiano Salcedo was responsible for the creation of most of the public gardens in Bogotá since 1881. Salcedo was responsible for the gardening of the city squares and became a local of the plant commerce business and the introduction of plants to Bogotá. In 1864, he and a man named Manuel Párra rented a colonial house with seven large patios, where they established the city's first greenhouse commercial nursery and from where they endowed the city with a completely new repertoire of plants.[34] Salcedo's name appears in the history of Bogotá in association with all works or any other activities involving plants during his tenure. For instance, he is identified as part of the group that organised the horticulture and floriculture section of the 1899 National Agriculture Exposition. For this event, besides his role as part of the organising staff, Salcedo presented some native ornamental plants such as begonias (*Begonia* spp.) and *Gesnerias* along with some other imported species without any botanical description (e.g. dwarf palms). His efforts won him the Silver Medal in the floriculture section judged by Fidel Pombo, Eugenio Pardo, Julio Arboleda, and Delfin Restrepo.[35]

The participation of a self-taught gardener in the creation of a distinctive urban nature in Bogotá represents a unique feature in relation to the cities in the south of Latin America. In cities better connected with the emergent international trade such as Buenos Aires or Rio, the economic flourish allowed local administrations to hire highly recognised Landscape Architects to undertake ambitious urban renovations. In these cities Frenchmen who had either worked close to Haussmann or been involved in the landscape designs for the international exhibitions hosted in Paris in 1867 and 1900 contributed to the creation of the new urban nature. Example of these are the works of Charles Thays in Buenos Aires, Santiago and Montevideo, Joseph-Antoine Bouvard's contributions in Sao Paulo and Claude Nicolas Forestier's in Buenos Aires.[36]

The Salcedo case is only comparable with the work of Latin American urban historiography has recently revealed the unusual participation of a self-taught gardener Horacio Arreondo in Uruguay. Rather than a professional gardener, Arreondo was a self-taught historian and archaeologist who after investigating the Hispanic fortress in the Santa Teresa Region (189 miles east of Montevideo) was appointed by Uruguayan President Baltasar Brum (president from 1919 to 1923) to undertake a huge transformation of this space into a natural park. As Salcedo, to accomplish his task, Arreondo used books. In his particular case, his work as gardeners was based on the works of Édouard-François André and Eugène-Emmanuel Viollet-le-Duc.[37]

Genaro Valderrama

The name Genaro Valderrama appears with frequency in documents of the early years of Bogotá's gardening history. As a young man, Valderrama showed a marked interest in botanical studies. In 1859, he participated in the creation of the Sociedad de Naturalistas Granadinos (Nueva Granada Society of Naturalists) where he received formal training from the botanist Francisco Bayón.[38] Valderrama can be considered the first person with botanical formation to develop a conscience of the importance of plants in the construction of urban spaces, as well as one of the first to recognise the ornamental potential of the Colombian flora. Valderrama's formal education shows his position as a member of the elite. His social position endowed him with spatial mobility. In a time when most of the people rarely left their city, he had travelled to Europe where he had direct contact with the park and gardens that Casiano Salcedo only knew through books.

Having travelled to Paris where he visited some of the parks,[39] to Valderrama's eyes, his experience endowed him with an aesthetic sensibility that made him the most qualified person to create and administer Bogotá's newly established Parks and Public Gardens Office. For this reason, during the first years of the production of green spaces, Valderrama assertively sought a leading role in the construction of the city's gardens and parks. For example, among the profuse correspondence he sent to the Ministry of Development, he presented himself in 1889 as the best candidate for the that-time vacant position of Parks and Public Gardens Administrator.[40] This proposal was rejected by the government, as some months later Valderrama presented another project aiming to revive the Expedición Botánica's small botanical garden beside the astronomical observatory.[41] This suggestion was also rejected by the Ministry arguing that the site was offered to the Engineering College for the construction of a new building.[42] Disappointed with the lack of response from the local administration, he wrote to the Minister complaining as follows:

> I have rendered important services to my country, among which outstands the fact of have been made known some of our spontaneous products in the overseas. . . . Because of this, I receive honours and awards in Europe as is proven in the documents—which I attach to be returned—. . . But in my homeland, no one has been grateful to me, and to the contrary, I have only received disdain and deceptions which cannot be expected in a civilised country.[43]

After the letter the Ministry did assign Valderrama a plot of land next to the train station headquarters to develop his plans; however, there is no information showing whether or not he undertook any gardening works in this space.

As a consequence of the government's nonrecognition of his qualifications, Valderrama was overshadowed by Casiano Salcedo and played a subordinated role in the construction of Bogotá's green spaces. Valderrama was a bitter critic of Salcedo's works, deriding their 'lack of merit' while continuing to present himself as the best person to create green spaces 'under the prescriptions of art'.[44] Although it is not completely clear whether or not Valderrama managed to obtain the position of Administrator of the city's green spaces, despite not achieving his goal of being directly in charge of the creation and modification of the city's gardens, he held the position of General Inspector of Parks and Public Gardens several times, during which he orchestrated an expansive campaign of moral hygienisation of those spaces.

Despite his formal scientific education, like Salcedo and many others that followed them, Valderrama was essentially a dilettante. Valderrama and Salcedo were the most prominent figures among a group of self-taught gardeners who emerged in Bogotá as a consequence of the absence of a gardening tradition and formal education in this area. Despite the differences between Salcedo and Valderrama, it is important to highlight that they both embodied a particular type of knowledge regarding the construction of an adequate urban environment. In other words, they were men of their time: although their creations aimed at an ideal of 'progress' that embodied nothing more than a desire to resemble Europe, they also represented particular ways to understand plants. As a consequence, the creation of Bogotá's modern urban green spaces resulted in a sort of hybrid that combined a wide range of influences as represented not only in the gardeners themselves but also in the plants they used in the creation of gardens and parks.

Notes

1 Jaime Jaramillo Uribe, "La visión de los otros. Colombia vista por observadores extranjeros en el siglo XIX," *Historia Crítica*, April 19, 2017, https://doi.org/10.7440/histcrit24.2002.01; Camila Torres Torres, "La imagen de Bogotá construida por los viajeros extranjeros que recorrieron el país a lo largo del Siglo XIX," MSc diss, Universidad Javeriana, 2009.

2 Amin and Thrift, *Cities: Reimagining the Urban*; Eugene McCann and Kevin Ward, *Mobile Urbanism: Cities and Policymaking in the Global Age, Globalization and Community*, vol. 17 (Minneapolis: University of Minnesota Press, 2011).

3 Guy Debord, *The Society of the Spectacle* (New York: Zone Books, 1994).

4 See, Torres, Rondy. "Tras las huellas armoniosas de una compañía lírica: La Rossi-D'Achiardi en Bogotá." *Revista del Instituto de Investigación Musicológica "Carlos Vega"* 26, no. 26, (2012): 162–200.

5 The architect Luis Fernando Gonzales accomplished a comprehensive review on the national exhibitions made in Colombia, see Luis González, *Del Alarife al Arquitecto: El Saber Hacer y El Pensar La Arquitectura En Colombia: 1847–1936* (Medellín: OjoXojO, 2013), 43–148.
6 *Papel periódico Ilustrado*, "Exposición Nacional," Bogotá, no. 2, Octubre 1, 1881, 32–33.
7 José Perdomo, *La Opera en Colombia 1817–1980* (Bogotá: Arco, 1979), 131.
8 Correspondence between the Bogotá's Gardeners Union and Chair of the National Public Works complaining about the public works undertaken by Abel Riaño, a non-member of the union, November 3, 1934, folio 74, tomo 287, fondo Ministerio de Obras Públicas, AGN.
9 Correspondence between José Gómez, Din of Mathematics and Engineer department of the Universidad Nacional, and Chair of National Building, September 17, 1936, folio 295, tomo 287, fondo Ministerio de Obras Públicas, AGN; Correspondence between (Signature illegible) Santamaria, Deputy Director of National Buildings management, and the Parks and gardens administrator, 23 August 1935, folio 166, tomo 287, fondo Ministerio de Obras Públicas, AGN.
10 Correspondence from Luis Bautista Manager of the Apulo hotel to Minister of Public Works, October 8, 1934, folios 64,65,69 tomo 287, fondo Ministerio de Obras Públicas, AGN.
11 The petition made by all of the city's gardeners to Minister of Public Works, April 24, 1936, folio 253, tomo 287, fondo Ministerio de Obras Públicas, AGN.
12 About the work of Thomson in Bogotá, see Claudia Cendales, "Robert Thomson (1840–1908): A Kew Gardener in Bogotá, Colombia," *Garden History* 40, no. 2 (2012): 239–52.
13 Correspondence between Robert Thomson and Minister of Development, January 30, 1893, folio 25, tomo 823, fondo Ministerio de Obras Públicas, AGN.
14 Correspondence between Robert Thomson and Minister of Development, January 30, 1893, folio 26, tomo 823, fondo Ministerio de Obras Públicas, AGN.
15 Correspondence between Robert Thomson and Mr A. Torres explaining Thomson's suggested changes for the city's parks, September 13, 1893, folio 44, tomo 823, fondo Ministerio de Obras Públicas, AGN.
16 From this point onwards, the money values represented as *pesos* refer to Colombian currency known as *pesos*.
17 Correspondence between Robert Thomson and Mr. A. Torres explaining his suggested changes for the city's parks, September 13, 1893, folio 44, tomo 823, fondo Ministerio de Obras Públicas, AGN.
18 Correspondence between Robert Thomson and Minister of Development, January 30, 1893, folio 28, tomo 823, fondo Ministerio de Obras Públicas, AGN.
19 Correspondence between Robert Thompson and Minister of Development, May 13, 1891, folio 17, tomo 823, Fondo Ministerio de Obras Públicas, AGN.
20 See, for instance, John Reader Jackson, *Commercial Botany of the Nineteenth Century : A Record of Progress in the Utilisation of Vegetable Products in the United Kingdom and the Introduction of Economic Plants Into the British Colonies during the Present Century* (Cassell & Co., 1890).
21 Norman Thomson, *The Putumayo Red Book*, 2nd ed. (London: N. Thomson & Co, 1914), 18–19.
22 Antonio Izquierdo, *Memorial Sobre Agricultura* (Bogotá: Escuela Tipográfica Salesiana, 1909).

23 Inés Sanmiguel, "Japoneses en Colombia. Historia de inmigración, sus descendientes en Japón," *Revista de Estudios Sociales*, no. 23 (Abril 2006): 83.
24 Izquierdo, *Memorial Sobre Agricultura*, 38; Antonio Izquierdo, *Riqueza Nacional: El Caucho* (Bogotá: Tipografía de la Gaceta, 1910), 75.
25 Daniel Ortega, *Apuntes Para La Historia de Chapinero* (Bogotá: Editorial Kelly, n.d.), 49.
26 Carlos Arturo Niño Murcia, *Arquitectura y Estado: Contexto y Significado de Las Construcciones Del Ministerio de Obras Públicas, Colombia 1905–1960*, 2da ed. (Bogotá: Universidad Nacional de Colombia. Unibiblos, 2003), 55.
27 Although regularly cited by scholars interested in the history of Bogotá, the gardening works developed by Kawaguchi have not been addressed in any serious, in-depth investigation. Thus, a series of questions regarding the non-European influence in the construction of urban gardens remain unanswered. For example, did Kawaguchi use native or introduced plants? How did he obtain the plants to create the gardens? Did he establish an Asian style in the public gardens he made in the city?
28 About the Tatsugoro Matsumoto's works in México, see Sergio Galindo, *Los que vinieron de Nagano: una migración japonesa a México* (México: Artes Gráficas Panorama, 2015).
29 Gabriel Ortiz Williamson, "Casiano Salcedo," *Cromos*, no. 106 (March 16, 1918): 139–49.
30 Some of the nineteenth-century gardening books found in the libraries of Bogotá are as follows: El Real Jardín Botánico de Madrid, *Semanario de Agricultura y Artes Dirigido á Los Párrocos*, XXIII vols. (Imprenta del Villalpando, 1808); M. Roitard, *Nouveau Manuel Complet de L'Architecte Des Jardins Óu L'art de Les Composer et de Les Décorer*, 9th ed. (Paris: La Librarie Encylopedique de Roret, 1830); Juan de Sandóval, *Nuevo Manual de Jardinería y Horticultura Ó Sea Tratado Complero Del Cultivo de Árboles de Ornato y Frutales, Flores y Hortalizas* (París and México: Libreria de la Vda de Ch. Bouret, 1897).
31 Cendales Paredes et al., *La Vida Privada de Los Parques y Jardines Públicos*, 73.
32 Ortiz Williamson, "Casiano Salcedo."
33 Cendales Paredes et al., *La Vida Privada de Los Parques y Jardines Públicos*, 73.
34 It seems that the nursery business was successful not necessarily because of the sale of living plants but rather, as Salcedo confessed, due to the sale of flower bouquets used in the ornamentation of churches attended by gente rezandera (lit. 'praying people'). Ortiz Williamson, 'Casiano Salcedo.'
35 Jurados de Calificación and Junta Organizadora, *Exposición Nacional de 1899. Catálogo de Las Diferentes Secciones* (Bogotá: Imprenta de Luis M. Holguín, 1899), 6, 90, 107.
36 Sonia Berjman and Anatole Tchikine, "Landscape Architecture in Latin America: Nineteenth and Twentieth Centuries," *Studies in the History of Gardens & Designed Landscapes* 39, no. 3 (July 3, 2019): 175–77, https://doi.org/10.1080/14601176.2018.1561817.
37 Alicia Torres Corral, "A Paradoxical Paradise: Parque Nacional Santa Teresa, Uruguay," *Studies in the History of Gardens & Designed Landscapes* 39, no. 3 (July 3, 2019): 199–212, https://doi.org/10.1080/14601176.2018.1491738.
38 Francisco Bayón was an important character in the history of Colombia's botanical development. Along with his teacher José María Cespedes, Bayón devoted considerable efforts toward the modernisation of botanical knowledge, importing books to explore sub-fields such as physiology and vegetal pathology as well as learning new approaches to understanding plant anatomy and taxonomical classification. Diana Obregón, "La sociedad de naturalistas neogranadinos y la

tradición científica," *Anuario Colombiano de Historia Social y de la Cultura* 0, no. 18–19 (January 1, 1991): 101–23; Santiago Díaz Piedrahíta, "La Escuela de Ciencias Naturales de La Universidad Nacional," in *La Universidad Nacional En El Siglo XIX. Documentos Para Su Historia, Colección CES*, ed. Estela Restrepo Zea (Bogotá: Facultad de Ciencias Humanas, Universidad Nacional de Colombia, 2004), 15.

39 Cendales Paredes et al., *La Vida Privada de Los Parques y Jardines Públicos*, 93.

40 Correspondence between Genaro Valderrama and Minister of Development, October 22, 1889, folio 5, tomo 823, fondo Ministerio de Obras Públicas, AGN.

41 The Royal Botanical Expedition to New Granada was a vast scientific enterprise undertaken by the Spanish Crown in the current Colombian territory between 1783 and 1916, see Daniela Bleichmar, *Visible Empire: Botanical Expeditions and Visual Culture in the Hispanic Enlightenment* (Chicago: University of Chicago Press, 2012); see also Pilar San Pío Aladrén, ed., *Mutis y La Real Expedición Botánica Del Nuevo Reyno de Granada*, 1da ed. (Bogotá, Colombia and Barcelona, España: Villegas Editores ; Lunwerg Editores, 1992).

42 Correspondence between Genaro Valderrama and Minister of Development, March 2, 1890, folios 9–13, tomo 823, fondo Ministerio de Obras Públicas, AGN.

43 Correspondence between Genaro Valderrama and Minister of Development, January 2, 1890, folio 8, tomo 823, fondo Ministerio de Obras Públicas, AGN.

44 Correspondence between Casiano Salcedo and Braulio Vélez, (Minister of Development), October 16, 1889, folio 5, tomo 823, fondo Ministerio de Obras Públicas, AGN; Correspondence between Genaro Valderrama and Braulio Vélez, Minister of Development, October 25, 1889, folio 6, tomo 823, fondo Ministerio de Obras Públicas, AGN.

6 The plants

Utterly unnecessary in the rural world, the figure of the gardener was the epitome of modern urban life. Gardeners emerged in the nineteenth-century Bogotá and mediated in the migration of plants from the private to the public sphere. Thus, after having exploring the creation of a urban nature in the guise of parks and gardens and having reviewed the gardeners involve in their construction, it is time to come back to plants. To explore the nature of the raw material used by gardeners, namely plants, the following pages address the socioenvironmental constraints and negotiations involved in the development of a *sui generis* urban flora that, in turn, made of Bogotá a floristic island richer in plants species than its surroundings. This section will reveal how the creation of modern Bogotá entailed the thriving coexistence of native and naturalised plants alongside the foreign plants introduced to the city adding a new botanical and ethnobotanical layer to the city's history.

The European image of nature

Similar to all the Latin American cities, Bogotá was a European project.[1] The political and social institutions, the religion, and the ways of understanding nature came from the other side of the Atlantic and underwent a process of integration with local expression and sensibilities. These encounters were also evident in the vegetal realm.[2] The conjugation of floras that place together native vegetables such as maize, *ullucos* and potato with introduced species such as apples, onion, or lettuce remained more or less stable until the end of the nineteenth century. After the independence in the early nineteenth century, the model to follow was not Madrid but London or Paris, the former worked as the economic model to follow and the latter as the cultural one.[3] From these new poles of 'progress', the emergent elites of Bogotá introduced the ways of building the city and its green spaces. This external influence limited the full development of local gardening practices to the extent that Andean species of an enormous ornamental potential

DOI: 10.4324/9781003208167-9

such as begonia and certain types of Ericaceae and Melastomataceae had a marginal role in the creation of gardens in Bogotá. The integration to the international trade underpinned this dependence when books, gardening ideas and most of the aesthetic universe present in the city came to keep flowing with few resistances from Europe.[4]

In Europe, the nineteenth-century industrialisation and massification of urban spaces meant a drastic transformation of the ways of being on the earth. Compared to the rural bodies of previous centuries, those that emerged along the urban spaces expressed a vastly different approach to the environment. Modern urban bodies had become exiled from the pre-industrial rural environments, and their nostalgia for their 'lost-paradise' was manifested in the emergence of romantic and illustrated depictions of nature that embodied the deep contradictions posed by modernity.[5] Starting with Alexander von Humboldt, the exaltation of nature in scientific terms led to the appearance of dozens of Europeans engaged in collecting, describing, painting, and later, photographing the natural richness of tropical America.[6] However, these visual appropriations of tropical nature also posed an enormous contradiction: however spectacular they might appear to the eye, the climatic conditions of these tropical lands combined with a myriad of insects, and exuberant plants and estrange mammals were not suitable spaces for civilised life. This environmental determinism would strongly impact the ways in which urban nature was built in the tropical cities in Latin America. Whereas the Europeans showed themselves overwhelmed by the tropical nature that led to a large representation of it in controlled artificial ecosystems (i.e. greenhouses), in Bogotá the model of nature followed by the Creole people was the native of Europe. The exaltation of temperate nature that, in theory would allow the Bogotanians to achieve a certain level of civilisation, was widespread and manifested in romantic and artistic expressions, in ways of teaching about natural phenomena, and in scientific ways of understanding the environment and biological organisms.

Colombian approaches to teaching about biological phenomena exemplified the dominance of the illusion of temperate nature. Guide textbooks that were first published in French and later translated into Spanish presented explanations of the mechanics of nature; however, all of the examples derived from temperate contexts. An article written by a Frenchwoman named Claudia Juranville in 1872 was published in the journal *La Escuela Normal, Periódico Oficial de Instrucción Pública* (Official Newspaper of Public Instruction). Devoted to demonstrating the benefits obtained with the study of botany, this article presented a hypothetical teacher trying to persuade a pupil named Amelia to undertake studies on plants. As the teacher is describing the behaviour of plants in different seasons, Juranville quotes someone named Keraboy to explain the reaction of plants during the spring:

> As soon as the sun of May has warmed the earth, when the leaves are seen to leave, to break or to reject the tunics that have served them as a cradle. Comb the trees with green hair under which their foreheads rejuvenate. . . . Innumerable leaves suddenly spread through the air, similar to the sword that comes out of the scabbard, the fan that opens, or the piece of cloth that unrolls. A few days pass and the groves have become covered too, and the shadow has darkened in such a way that one is tempted to ask where those rich and fresh draperies that the mansion of the human race has been decked out in had been kept.[7]

The phenomenological experience of the spring manifested on trees would have been almost impossible to understand for people living in the ubiquitously green Andean ecosystems.[8] Thus, teaching the biological process through the lens of European temperate nature was the reflection of a social construction of nature that in many cases did not at all match with the environment in which the people of Bogotá were immersed.[9]

The vast romantic literature produced by the republican generation of *men of letters* reinforced the exaltation of temperate natural phenomena. Notably, the Romantic movement emerged in Bogotá along with other modernist features, and, as highlighted by Charles Safray, its thriving development was reflected in the Bogotanians' proliferating productions of poetry:

> Bogotá is the most civilized city in the Nueva Granada. In addition to the classic studies, the medical and law studies have been imparted. However, the vocation for science is less generalized than the love for the letters. The liberal arts do not thrive here although the people from Bogotá love everything that is related to them. It could be affirmed that all of them are born being artists or poets. As there is no need to learn poetry, they are superior in this art than in others, and although Nueva Granada's most outstanding poets do not reach a level of brilliance it cannot be denied their inspiration.[10]

The most commonly used metaphors and similes around nature referred to forests, seasons, flowers and other elements of temperate landscapes, and despite the lack of any winter or spring seasons, nineteenth-century Colombian poetry is full of descriptions about these climatic phenomena. For instance, José Asunción Silva, one of the most important nineteenth-century Colombian poets, wrote: 'He was a great lyric poet that speak to the earth in a winter afternoon . . .'[11] Another Colombian Poet, José Eusebio Caro wrote a paean to the Cypress tree as follows: 'Yes, funeral cypress/when the night, when its quiet shade surrounds you/when hidden, the alone owl in your branches flap/my dead father shadow appears wandering among your lefts'.[12] As is revealed in Caro's poem, temperate trees and fruits such as apples were more common than for example, native fruits such as the banana passion-fruit or the *guanabana* (*Annona muricata*). The exaltation of the own nature would not appear with any force until the first half of the twentieth century.[13]

However, the dystopic lens from which the tropical nature of Colombia was explained, taught and versed was spotted by some who advocated the recognition of their own nature. In 1886, an author using the pseudonym of *Mindrelio* published in a local newspaper a satirical narration about an annoying literature critic of Bogotá. In the voice of the main character, the author says:

> You must know my friend, that one of the capital fails committed by those that write here, is not to use analogies or similarities (which are the soul of comparisons) in nature and facts observed by themselves. *They take everything from the European books*. It makes you laugh, being in Bogotá or Medellín, to listen to talk about the nightingale or the larks' sings, something that they never have heard. Or hearing about the unjoyful morning of April, which might be as splendid as you want in Europe or in temperate zones. Because here they are the saddest, the cloudiest, the most inclement and humid ones. In a cold and insensible state is the heart when our man of letters compares any common thing with the Spring, Autumn or any other natural phenomenon from other climates, of which we only have known through the books of geography [emphasis added].[14]

The ubiquitous presence of temperate nature in Bogotá's cultural context fed a general perception that plants native to temperate regions were most suitable for the construction of the modern urban green space. The modern city not only followed the ways of building and living expressed in Paris or London, but it also sought to use the same types of photosynthetic raw materials in the construction of its gardens, parks and urban forests.

European plants in the creation of Bogotá's green spaces

The exaltation of temperate plants as the most suitable elements in the construction of green spaces in Bogotá was initially promoted by European experts. Embodying their own native conceptions of plants, they highlighted the pertinence of using European species as raw materials in the construction of gardens and parks. For example, in the *Tratado Elemental de Agricultura y Ganadería*, the Spanish José María Gutierrez de Alba enumerated the trees that could be used to enrich and revegetate places destroyed 'by man's hand' in Colombia:

> About the resinous trees, (I recommend) pines of different species. Among them, it could be chosen those most suitable types for each weather and soil conditions, from the sandy grounds in the coast to the mountain's summits. Amongst the non-resinous trees, it could be acclimatised the cork tree [*Quercus suber*], the holm oak [*Quercus ilex*], the European nettle tree [*Celtis australis*], the beech [*Fagus* spp.], the white and black poplar [*Populus alba* and *P. nigra*], the birch [*Betula* spp.], the ash [*Fraxinus excelsior*], the plane [*Platanus* x *hispanica*], the horse chestnut [*Aesculus hippocastanum*], the elm [*Ulmus procera*], the linden [*Tilia* × *europaea*], and many thousand others.[15]

All of the trees suggested by Gutierrez de Alba corresponded to species with large symbolic significance in Europe. It can be seen that he essentially extrapolated his own understanding of nature and tried to apply it in a completely different context. When describing suitable plants species to be planted in the Colombian cold weather regions (such as Bogotá), he again recommended a group of common European trees.

> Which are the most suitable trees for cold weather places in Colombia? Understanding for cold weather places where the thermometer never goes under 7 or 8 °C above zero. Being those temperatures here the most common ones. . . . It is possible to cultivate with success many of the temperate-region trees, especially the common cherry, several species of plums, the raspberry [*Rubus idaeus*], the gooseberry (*Ribes uva-crispa*), the apple tree, all the species of the *Persicum* genus such as the peach tree, the freestone peach. Also, some other trees such as the walnut [*Juglans regia*], the chestnut [*Castanea sativa*], the hazelnut [*Corylus avellana*] and many other could be acclimatised.[16]

Additional suggestions around the suitable plants to be used as part of the city's ornamentation were made by the British gardener Robert Thomson. Although the plant species recommended by Thomson were not necessarily native to Europe, they were commonly used as ornamental plants in London, the city where he had lived and learnt the gardening arts. In his proposal for plants to be incorporated into the Parque de Bolívar, he wrote:

> If this garden was deprived of the few pine trees which adorn it, there would be hardly anything left worthy of so important site, for nearly all the other plants therein are very common and ill-adapted, as well as asymmetrically planted. To establish this garden in accordance with its position,

> selections of fine foliaged and striking plants should be a very crucial feature. *Some of the finest plants in Europe gardens would find a fitting home in this as well as well in other parks of Bogotá* [emphasis added].[17]

In the same letter, Thomson suggested:

> For the adequate planting and renovation of these parks in the manner indicated it is indispensable to import and to propagate numerous species and varieties of the choicest flowering and decorative temperate plants of European gardens. For instance, many species of splendid foliaged Palms from Australia and other temperate regions; rhododendrons [*Rhododendron catawbiense*], azaleas [*Rhododendron simsii*], heaths [*Erica carnea*], epacris [*Epacris* spp.], magnolia [*Magnolia* spp.], etc., etc. to be planted in masses, and of the seasons and flowering. Also, numerous varieties of other choicest flowering plants some of which are unknown here, these to be cultivated in large quantity.[18]

The plants suggested by Thomson to be used in the gardens of Bogotá can be analysed from a botanical perspective. For instance, he mentions the convenience of using rhododendrons, heaths, and epacris, all members of the broadly distributed Ericaceae family, of which 287 species are native to Colombia and some have ornamental potential.[19] However, perhaps due to a lack of knowledge concerning Colombia's botanical richness, none of these native species were considered by Thomson as suitable elements in the ornamentation of chilly Bogotá. Rather, he suggested alien members of the Ericaceae family such as rhododendron, heath and epacris with which he was acquainted as part of his gardening work in London. However, Ericaceae family species from the Andes Mountains such as *Carbonero* (*Bejaria aestuans*), *Gaultheria erecta* and *Cavendishia* spp. have a great potential as ornamental.

People and institutions involved in the plant importation

Whereas foreigners like Robert Thomson or Gutierrez de Alba played key roles in the strengthening of the historical predilection for the European species, they did not significantly participate in the active introduction of plants from Europe and the resulting physical representation of the European urban flora in Bogotá. Rather, it were locals who ruled the creation of this new urban nature. They undertook a remarkable process of importation and acclimatisation of exotic plants in their efforts to recreate a temperate nature. As social actors historically placed in a time of trade expansion, they managed to build a relatively solid network of plant suppliers. Through these suppliers, a considerable number of new species of plants, not only from Europe but also from other tropical regions (e.g. Asia), arrived in the city. These new sets of *not seen thus far* plants further enriched the already mixed flora of the city.

By the last decades of the nineteenth century, the social relevance of gardens and parks was undeniable. Turned into raw materials in the construction and constant maintenance of these places, plants became products linked to the market's logic. Consequently, securing the supply of plants was among the main concerns among their administrators. Given the endemic economic constraints of the city, local administrations faced continuous problems around the plant supply. As a consequence, the sometimes-precarious state of parks was a recurrent source of complaint in Bogotá. As a practical solution, the municipality passed the problem on to private citizens, who oversaw public tenders in park administration and maintenance.

However, transferring the management of parks and public gardens to private hands did not solve the problems around the plant supply. There was a lack of a well-established local market to obtain the required plants due to the absence of nurseries to grow the plants. Consequently, many of the plants used in the creation of the modern green space in Bogotá arrived via donations. These difficulties explain the relatively poor diversity of plants initially used in the construction of green spaces. For example, Casiano Salcedo, the director of the city's Parks and Public Gardens Office, reported the plantation of only nine different types of plants during his two-year tenure from 1892 to 1894.[20] It is interesting to point out how under the precarious economic milieu, Salcedo primarily planted already popular ornamental plants, which, with the exception of the amarrabollos (*Meriania nobilis*) and fuchsias (*Fuchsia* sp.), were temperate species historically used in the patios and solares of Bogotá (see Table 6.1).

Table 6.1 Some of the plants planted in parks and city gardens between 1892 and 1894

Plants	*Scientific name*	**Origin**[21]
Araucarias	*Araucaria heterophylla*	Australia (Norfolk Island)
Amarrabollos	*Meriania nobilis*	Southern America (Andes)
Acacias	*Acacia* sp.	Australia
Roses	*Rosa* sp.	Asia temperate (but now a cosmopolitan)
Lilies	*Lilium* sp.	Europe and North America
Carnations	*Dianthus caryophyllus*	Europe Mediterranean (probably)
Fuchsias	*Fuchsia* sp.	Southern America (Andes)
Hortensias	*Hydrangea macrophylla*	Asia temperate
Heliotropes	*Heliotropium arborescens*	Southern America (Perú)

The use of these few plant species would continue with his successors. Once Salcedo left his position as a contractor, he was replaced in 1894 by Agustín Torres, who obtained the management of the city's parks and public gardens after presenting a self-redacted contract to the Ministry of Infrastructure. In the deal, Torres promised to undertake all the necessary actions to maintain parks in optimal conditions. For instance, in clause two he agreed to 'keep in the parks a permanent and variated planting of flowering plants, bushes and trees planted with art and symmetry'. In addition, he guaranteed that he would 'plant elms and pines among other foliage trees instead of the enormous eucalyptus currently existing'.[22] Again, as we can see from Torres' declaration, the notion of parks is associated with the very limited number of commonly used, mainly European tree and plant species.

After Torres' apparently disastrous management, the administration of parks and public gardens sought, once again, the services of Casiano Salcedo. In the proposed contract between Salcedo and the Ministry of Infrastructure, published in the *Official Diary* on the 4th of July 1896, the government promised to pay $440 per month for the administration of the city's parks and public gardens.[23] However, this contract was never signed. It seems that the government was reluctant to commit to such a large budget, and they decided to open a public tender to find another contractor. On the 18th of July 1896 (12 days after the first contract), Fernando Faverna won the tender with the minimum offer of $320 Colombian pesos [pesos hereinafter] per month (later renegotiated to $310) and assumed the position of administrator of Bogotá's parks and public gardens.[24]

There are some interesting clauses in the contract signed between Faverna and the Ministerio de Hacienda (Ministry of Finance), which was the same as that proposed for Salcedo with the exception of the salary. First, the government allowed the contractor to introduce the plants that 'he considered convenient for their ornamentation'. In the same vein, the contractor could 'sell, according to his will, the flowers produced in the parks for the ornamentation of public buildings'. The first clause (which was also included in a large number of later contracts) allowed the parks and public gardens manager to impose his particular taste, thus revealing the government's lack of attention to the plants' significance in contributing to the city's image. Despite the rich discourse about the role of plants in the city, *plants* were a very general concept without any real form for the city administration. Thus, contrary to what would happen some decades later, plants were not yet undestood as symbols of national identity.[25] Equally, the contractor's full control over the flower choices reveals both the continued lack of resources to care for those spaces and the Ministry's tremendous ignorance of the significance of gardens: to the local government, the flowers in the parks were merely goods to be used in the decoration of the city's public buildings.

In addition to his inventory of the plants present in the green spaces of Bogotá at the end of his 1894's tenure, Salcedo made a report in which he illustrated the physical conditions of those spaces, including the trees and plants used in their ornamentation (see Table 6.2).

Table 6.2 Most of the trees present in Bogotá's green spaces in 1897

Green space	*Tree*	*Number*
Plaza de Bolívar	Pines	19
	Araucaria	2
	Quinos (*Cinchona* sp.)	4
	Ethiopian banana (*Ensete ventricosum*)	2
	Siete cueros (Tibouchina lepidota)	2
	Amarrabollo (Meriania nobilis)	2
	Patula pine (*Pinus patula*)	3
	Pine, planted by Mutis.	1
Jardín del Observatorio	Other big and small pines	
	Peruvian pepper (*Schinus molle*)	1
	Duraznillo (*Abatia parviflora?*)	n/s (not specified)
	Acacias (Acacia *sp*.)	n/s
	Amarrabollos	n/s
	Few and abandoned flowering plants	n/s
	Ancient Colombian walnut (*Juglans neotropica*)	1
Jardín de San Carlos	Ethiopian banana	n/s
	Roses	n/s
Garden in the Plaza de San Carlos	Pines	4 alive, 1 dying
	Gorse (*Ulex europaeus*)	n/s
	Sietecueros	n/s
	Abutillon o amarrabollo	n/s
Parque Santander[26]	Eucalyptus	8
	Amarrabollos	2
	Sietecueros	4
	Aralia (*Aralia* sp.)	1
	Cañabravo (*Arundo donax*?)	1
	Pines	Some (one is dry)
Parque Centenario	Many trees dying and getting dry. The flowering plants and flowers in a state of complete abandonment and disaster.	n/s
Parque Camilo Torres	Fenced with bare wire	n/s
Plaza de Los Mártires 'The trees need to be trimmed. There are few and common flowers. Full of weeds, its general state is of abandon'.	The best trees are the pines and the *Alcaparros*	n/s

Salcedo's report reinforces the limited flora used in the ornamentation of Bogotá's public spaces. It is interesting to highlight the generalised use of pines, which were introduced into the city in colonial times, and how along pines and eucalyptus, the records show the use of some native trees of the *melastomataceae* botanical family, namely *sietecueros* and *amarrabollos*. The latter, an easy-to-grow and flashy-floriation tree native to the mountainous Antioquia region, unveil an internal exchange of plants between different Colombian regions. Howevere, the use of these species could be considered an initial appropriation of the best-known native plants as a means to cover the lack of a local plant market. Nevertheless, the number and variety of plants and the physical condition of the parks presented by Salcedo's report testify to a complete state of abandonment in the city's green spaces, which is explicitly reinforced through additional commentary. As Salcedo explained the issue:

> In general terms, the state of those places of decoration and health is of careless and abandon. The labours in charge of the works have informed that the reasons for the current state of those places is the absence of payment for their works since the budget devoted for this has been totally expended.[27]

Faverna had to face several difficulties during his tenure in charge of park and garden administration. For example, he found that one of the employees was living in the gardening tools storage shed in the Parque Santander. Moreover, according to the inventory made by Salcedo, some necessary tools were missing.[28] However, when Genaro Valderrama, who was then the General Inspector of Parks and Public Gardens, checked on Faverna's progress in October of the same year, he reported that despite the paucity of tools, some important works had been undertaken in the city parks.[29] Concerned about the lack of resources to fulfil what had been signed in the original contract of July 1896, Faverna asked for an extension to his contract on the 2nd of April 1897. However, this new contract presented different terms. Faverna expressed the impossibility of undertaking all of the works he had initially promised to undertake in the initial contract of 310 *pesos* and requested that the monthly payment be increased to 400 *pesos*, the same value as the initial contract made with Casiano Salcedo, which was also similar to the monthly budgets offered in other public tenders. As a consequence, the Ministry decided to cancel the contract with Faverna and gave the position back to Casiano Salcedo.[30]

In April of 1897, Casiano Salcedo signed a six-month contract with the Ministry of Finance. Varying similar drafts of this contract exist, the only difference between them being the price, thereby documenting a sort of bargaining process between the parties. Whereas the government offered

500 *pesos* per month, Salcedo asked for 700 *pesos*. Ultimately, despite the apparently limited resources, a final monthly payment of 730 *pesos* was agreed between the parties.[31] With this considerable increase of 410 *pesos* in the monthly operation of Bogotá's parks and public gardens,[32] Salcedo undertook the most important planting process that had occurred in the city thus far when in less than six months, he planted 29,928 plants.[33]

An important question arises regarding Salcedo's campaign of plant enrichment: given the lack of a solid local business exploiting the native flora in its ornamental potential, what was the origin of all of these ornamental plants? To answer this question, it is essential to understand that despite the high freight prices accruing from Bogotá's extreme isolation, the city had managed to be integrated into the international trade networks. Such integration, which can be considered the cornerstone of the modernisation process, brought goods and European tastes to the city's colonial streets, albeit most goods reached the city via beasts of burden until late in the nineteenth century. The materiality from which Bogotá's modernity was built was displayed in the city: musical devices, perfumes, dresses, cars, and many other elements came to Bogotá through the activities of the so-called *casas comerciales* (commercial houses), institutions comprised of a commissions agent, sellers and intermediaries that were in charge of importing innovations produced overseas to the city.[34] Along with banks, the ever-increasing number of commercial houses during the late nineteenth century reveals that, however, remote and difficult to reach, Bogotá was the final destination for many goods.

Plants were among the commodities introduced to the city through commercial agencies. At least one commercial house in charge of importing plants had a branch in Colombia. Rochester's Live Plants Company had offices in Bogotá and Medellín and was responsible for the introduction of a significant number of new species into the city. In this regard, an advertisement published by this company in Medellín's newspaper offered its services as follows:

> The people that desire to buy fruit and ornamental trees, as well as any class of bushes and rare flowers from diverse world regions, in large or small quantities, cultivated in the enormous nurseries in Rochester, New York. . . . It is offered to each of the buyers an agriculture, horticulture and floriculture manual which is quite clear in explaining the cultivation the fertilisation and the trimming of plants.[35]

Live Plants Company was interested in becoming the official supplier of plants for Bogotá. It seems that the company's agent in Colombia, Mr G.W. McLane, successfully lobbied the Ministry of Finance, as the minister ordered Salcedo to choose the plants to be used in his revitalisation plans from Rochester's catalogue. However, Salcedo argued that given the business model presented by Mr McLane, would be overly expensive and complex, as it was based on the acclimatisation of plants before their introduction into the parks and garden. Although Salcedo did choose some plants offered by the Rochester company, he stressed that 'those plants are very expensive compared with the prices of the *ones brought from Paris*' [emphasis added].[36] This comment is extremely relevant, as it is the only proof we have about the direct importation of plants from France during the end of the nineteenth century. Thus, not only were the parks' layouts inspired by the European garden, but in at least this case, the raw materials involved in their construction were also labelled as coming from Paris.

The French nationality of some of the city's plants reveals how after the bloody divorce with Spain, even nature was inserted in the reorganisation of models followed by the Latin American nations. Spanish heritage was negated and replaced with the exaltation of England and (even more so) France, which became the model of modernity and excellence.[37] Architecture, dresses and social manners were imported from Paris. However, the importation of French nature was not new and did not begin with Salcedo; the virtues of plants from Paris were extolled in Bogotá's history as early as 1830, as evidenced by a notice in the city newspaper attesting that:

> Sr. Pedro Feliz and bother have just arrived from France with a complete selection of fruit trees, legumes seeds and flowers of any class. They offered them for a moderate price in their living house in Carrera Street number 23.[38]

It seems that most of the plants used by Salcedo in this refurbishing plan were effectively introduced from France. In a letter sent by the Rochester Company's new agent, Mr Byron Fish, to the Ministry in January of 1898, the North American mentioned that a large number of ornamental plants coming from New York were already known by the government, and that a previous attempt to close the negotiations had failed due to 'the inconvenient way in which the business was posed'. Fish proposed to skip the acclimatisation process and plant the plants directly into the parks, and he also offered a professional gardener supplied by the company to conduct the process.[39] Although the nature of this business makes it difficult to track its full dimensions in the urban landscapes of Colombian cities, this letter reveals that plants were a good attached to the trade networks in the same manner as many other material elements of modernity. The very existence of this kind of commercial house shows Colombia's predilection for foreign plant species while also clearly evincing the underdevelopment in the plant ornamental sector in a tropical country endowed with exceptional floristic richness.

Despite Salcedo's commitment to improve the state of Bogotá's parks and public gardens, he resigned after only five and a half months of his second administration. As he explained his decision in a letter to the minister, he lacked the appropriate tools to achieve his work and complained about the lack of support offered by the police department in the parks' surveillance as well as the lack of money to pay the workers under his direction.[40] Salcedo also described the constant problems related to the water supply:

> The Parque Centenario lacks water, I cannot use the water from the public fountains because the neighbours stopped me arguing that water was used by them in the domestic task. [Also] The waterpipe from the San Diego highs which initially supplied water to the park was destroyed by the beer company and the washerwoman.[41]

Salcedo's words highlight the severe economic constraints faced by the park and public gardens administration during the first decades of its existence.

Following the end of the contract between Salcedo and the Ministry, he fought a two-year legal battle to be reimbursed for the funds he had invested to purchase the plants planted in the park. Through his proxy, Augusto Torre, Casiano Salcedo claimed the payment of 7,529.95 *pesos*; however, the minister countered with an offer of only 3,000 *pesos*, which he attributed to the 'current economic constraints' and the 'elevated price paid by Salcedo in the bought of the plants'.[42] Although the outcome of this litigation was not registered in the archives, the situation highlights that although the lack of local gardening traditions might have prevented Salcedo from achieving proper European gardens (as was highlighted by Robert Thomson), the lack of political will expressed in the ubiquitous tiny budgets for that enterprise made the development of such knowledge largely untenable, with the resulting consequences reflected in the appropriation and use of plants.

The city's commitment to carry on with adding plants to its gardens and parks continued after Casiano Salcedo's tenure. In an 1897 contract made between José Leocadio Camacho, chief of the 5th section of the Ministry of Finance, and Gabriel Peña Solano, the latter committed to planting 50 *manzanos de Duitama* [*non-determined species*] trees in the Parque Centenario and Santander and the Plazas Bolívar and Los Mártires.[43] Although it is not clear whether or not those trees were planted, it is important to highlight the general homogeneity of the species mentioned in this contract, which contrasts sharply with the state of things during Casiano Salcedo's administration.

Finally, the problems that undermined the state of Bogotá's gardens and parks were not resolved during the 1890s. The economic difficulties that engendered the administrators' lack of well-developed gardening skills led to repetitive complaints about the general state of the city's gardens and parks. The situation was exacerbated by the country's largest civil war, known as *La Guerra de Mil Dias* (One Thousand Days War) from 1899 to 1902, which had deleterious effects on the city's infrastructure including the parks. The severity of the situation is reflected in a letter the Colombian Astronomer Julio Garavito sent to the Ministry of Finance in May of 1901, in which he stated, 'Respectfully, I am asking Your Honour to send a peon to weeding the Jardin del Observatorio which was abandoned because of the war'[44] Thus, the consequences of the warfare completely reversed many of the earlier efforts devoted toward creating the city's public green spaces.

Following the *Guerra de los Mil Días*, the Colombian economy finally began to improve. The Panamá excision orchestrated by the United States led to an important compensation of USD$25 million from this government, which was largely invested in infrastructure and boosted the exportation of coffee, leathers, bananas, oil, and other commodities. Colombia's long-delayed integration into the global economy as a supplier of raw materials led to considerable economic growth, as evidenced by an annual economy growth rate increase of 3.7% between 1906 and 1926.[45] The country's economic activation enabled the creation of a strong bourgeois class composed of socially recognised men characterised by their civil and religious commitments.

One of these men was Antonio Izquierdo. Commonly recognised as an important figure in Bogotá's burgeoning real estate business, Izquierdo was one of the wealthiest Colombians of the early twentieth century. He had an active role in the development of agriculture and (self)published several books addressing the potential presented by some tropical products to be inserted into international trade. Izquierdo presented himself as a defender of agri-business based on certain products such as rubber or cotton cultivation.[46] However, his publications were not the result of knowledge or academic research but rather a product of his status as a wealthy and influential man interested in agriculture topics and immersed in politics. Izquierdo was named by Rafael Reyes' government to undertake an official visit to the United States, Europe and the Far East to gather information and learn the agricultural practices developed in those regions.[47] This accumulated experience along with his interest in foreign plants made him one of the most important introducers of useful and ornamental plants to Colombia in general and particularly Bogotá. As he noted in an interview for a local newspaper published by his friend José María Vesga y Avila, his official contacts with the United States Agriculture Department had resulted in the remittance to the Ministry of Infrastructure of more than 2,000 plants of cold and hot weather suitable to be acclimatised in Colombia. Izquierdo also explained that

> equally, I sent several types of ornamental trees for the parks of Bogotá and a box with grass seed given to me by the North-American government, along with a *wardian case* with many rare plants. Regarding the grass seed, I must say they were recently introduced to the United States due to its proved quality.[48]

In addition to the seeds he imported from the United States, Izquierdo sent Colombia several seeds of grasses, fruit trees, grains, potatoes, eastern tobacco and flowers from England,[49] as well as live olives from Italy, which he thought would be suitable to cultivate in the dry regions of Villa de Leyva, and one ton of cotton seeds and dates from Egypt. From Ceylon, Izquierdo sent a magnificent collection of plants in wardian cases, of which he claimed, 'you should not think I am exaggerating, because I sent the most beautiful and interesting representation of what is known today around useful, fruit and ornamental trees'. Those plants made part of a larger cargo that included 10,000 small cuttings of *hevea* rubber trees,[50] 50 pounds of tea seeds and several boxes of cacao seeds.[51]

Izquierdos's travels and plant procurement efforts highlight how the introduction of plants was undertaken as part of the interwoven global trade networks of the early twentieth century. Izquierdo published a booklet in 1918 that revealed an incredible network of contacts that included over 20 nurseries in 11 countries (see Table 6.3).[52] Izquierdo obtained many different kinds of trees, which in many cases ended up becoming part of Colombia's urban and rural landscapes. It is not an exaggeration to present Antonio Izquierdo as one of Colombia's most important plant introducers during the twentieth century.

Antonio Izquierdo was part of the Sociedad de Mejoras y Ornato (SMO; Society of Improvements and Ornamentation). Founded in 1917, the society took over the responsibilities previously undertaken by the Sociedad de Embellecimiento de Bogotá (SEB; Society for the Embellishment of Bogotá), which ended in 1913 after approximately 14 years of operation. As in the case of its predecessor, the SMO was composed of elite men charged by the national government to oversee the city's material improvements, including its parks and public gardens, and some of its members shared Antonio Izquierdo's interest in exotic species. Divided into several commissions addressing particular areas of improvement, the Parks Commission was headed by José María Saíz and José Joaquín Pérez. This commission regulated and oversaw all activities related to the city's green spaces, including initial planting and ongoing maintenance. One task in which the commission was particularly

Table 6.3 Worldwide nurseries where Izquierdo had contacts [With the original spelling in Izquierdo's publication][53]

Country	*Nursery*
United States	St. Louis Seed Co., St. Louis
	Edmun D. Sturtewart, Hollywood, California
	J.M. Thorburn & Co. Seedman, New York
	George C. Roeding, Fancher Creek Nurseries, Fresno, California
	The Barteldes See Co., Lawrence, Kansas, Denver, Colorado
	Theodore Payne, 345 South Main St., California
	The San Dimas Citrus, Nurseries, San Dimas, California
	Luther Burbank, Santa Rosa, Sonora Co., California
	Vaughman See, Chicago 84–86, Randolph St.
	German Seed Plant Co., 326–30, South Main St., Los Angeles, Cal.
	J.G. Hartison & Sons Nurseries, Berlín, Md.
England	Fidler & Sons, Royal Berkshire, Seed Stores Reading
	Perry Hardy Plant Farm, Enfield, Middlesex
	H. Cannel & Son, The Nurseries, Eynsford, Kent
	Carters, Tested Seed, Raynes Park London
	One & all seed, Warehouse, 92 Long Acre, London.
	Miss C. M. Dixon, Elmcroft Nursery, Edinburgh, Kent
	C. Engelmann, Essex, England
	J.T. Wesr, Towers Hill Brentwood, Essex
	Carter Page Co., 52–53, London Wall, London E/C
	Stuart Low Co., Royal Nurseries Bush Hill, Park Madd
	Keynes William & Co., The Nurseries Chingford, Essex
	G.F. Letts, Seedman, The Nurseries Hadleigh, Suffolk
	Thomas S. Ware Ltd, Telham, Middlesex
	Hobbies, 17 Road St. Place
	Benjamin R. Cant & Sons, Colchester
Australia	F.H., Brimming, 64 Elizabeth St., Melbourne
Italy	Fratelli Ingegnoli, Corso, Buenos Aires, número 54, Milan
Germany	Haage & Schmidt, Erfurt
Spain	E. Veyrat Hnos., Valencia
France	H. Valtier, 2 Rue St. Martín, Paris
Ceylon	J.P. William & Brothers, Heneratgoda, Ceylan
	H.D. Constantine & Sons, Kalutara
India	W. Kuddrappa & Sons, Lalbagh Road Bangalore, India
Japan	Yokohama Nurseries
Asia- Malayan Federal States	J. de A. Pereiras, The Singapore & Straits Nurseries, Singapore

active was obtaining seeds from the foreign species they deemed suitable to be used in the city's ornamentation.

Being recognised men from the elite class, Saíz and Pérez had access to the highest spheres of political power, and they engaged in regular communications requesting foreign seeds from many Colombian consuls and ambassadors. Some examples illustrate this relationship between the country's elites and the introduction of urban ornamental plants. For instance, in 1918 the Colombian consul in Bilbao (Spain) sent the society acorns from a European oak tree (*Quercus robur*), which the diplomat claimed was a 'venerable tree in Spain because in its shadow the ancient members of parliament gathered to create the laws which ruled the Basque people'.[54] Commonly called English oak, this species is native to most of Europe west of the Caucasus and is perhaps one of the most important trees in European culture.[55]

It is critical to stress how following the establishment of the SMO, the city's seed sources also extended to other Latin American countries. In April of 1917, the society charged one of its members, José María Samper, to

> ask the diplomatic body of Colombia in Quito, *through telegraph*, to ship seed or cuttings of the tree used there, in Buenos Aires, Santiago and Montevideo named *Platán del Pichincha* [*non-determined species*] and, if it is possible, some information about its cultivation.[56]

As in the case of Izquierdo's travels, the efficiency of international trips and the communication innovations that made it possible to cut across distances brought about a redistribution of plant species.[57] In the case of the *platán de Pichincha* tree, it is especially interesting how the Society specifically ordered to Samper the use of the telegraph to request the seeds, thereby avoiding this member's physical displacement to the neighbouring country of Ecuador. The enhanced conveniences offered by these new developments are reinforced by an observation made by Alexander Von Humboldt in the early years of the nineteenth century: 'There is nothing as difficult to solve as the issues posed by the migrations of the plants useful to men, especially since the communications have reached this level of efficiency in all the continents.'[58]

As indicated earlier, Colombia's diplomatic relationships with other Latin American countries made it possible to obtain various kinds of seeds. In December of 1918, the SMO contacted Eduardo Urdaneta, the General Consul of Colombia in Venezuela, who was also in charge of the Colombian consulate in Mexico City, to request a shipment of seeds or cuttings of the tree named 'thunder' [*non-determined species*].[59] Although it is not clear whether or not the Society received the thunder tree seeds or cuttings, they received numerous other samples. For example, by the end of 1918, the society had obtained 43 different kinds of seeds from the Granja de Agricultura de Toledo in the Canelones District of Uruguay, which were sent by that country's Ministry of Industry. Notably, as part of the thank you note to the Uruguayan government, the society promised to send a cargo of Colombian tree seeds, which suggests the existence of a thus far ignored early twentieth-century plant exchange between Latin American governmental authorities and civic societies of different countries.

Finally, the society members did not seem to have any real concerns about the origins of the trees used in the city; rather, their decisions were motivated by a sense of pragmatism. For example, when the District Government requested the SMO's opinion about the most convenient trees to be planted in the Avenida Colón in 1918, The commission of parks composed by Saíz and Pérez replied that

> the commission thinks that the already planted trees in a good shape planted in the Avenue must remain. For the new trees to be planted must be chosen those that have been thriving in the city, mainly acacias (like the ones planted in the Boyacá Avenue), nogales and pimientos [*Schinus terebinthifolius*].[60]

The first of these species was a broadly used ornamental tree from Australia, the second was a native tree worshipped by the Muiscas, and the third was a tree native from the north of Argentina. This pragmatic position held by the Society led them to create the first nursery to endow the city with locally produced trees using either native or species from all around the world (see Figure 10.5).

The native species in the construction of green spaces

Despite the overwhelming importance of temperate plants in the development of Bogotá's modern green space, it would be unfair as well as ecologically impossible to claim that the production of gardens and parks was entirely supported by the use of exotic and introduced plant species. Although less important than the European trees, native species also had a historical role in the creation and transformation of the city's green spaces. As previously presented, native willows were used in the production of the hedges for the purpose of plot delimitation in Bogotá's outskirts. In the same vein, early in the planting process, flashy blooming trees such as *amarrabollos* (*Meriania nobilis*), *siete cueros* (*Tibouchina lepidota*) or *alcaparros* (*Senna viarum*) were used. In this regard, the presence in Bogotá of the *amarrabollo* tree, a native species growing in the mountains of the Antioquia region, exemplifies how the city not only lured people from different regions but also attracted plants from other Andean ecosystems. Similarly, the role of Bogotá as a pole of attraction for other provinces' floras is exemplified by the efforts made by the SMO in 1918 to obtain seeds of the trees known as *manzano* and *santacausa* [*non-determined species*], both of which were species from the Sierra Nevada Mountains close to the Caribbean city of Santa Marta.[61]

In addition to Colombian walnuts, willows and *cerecillos*, a few other local trees were deemed adequate to be planted in the city. For instance, in 1899, the Sociedad de Embellecimiento de Bogotá (the institution that preceded the SMO) invited well-recognised Colombians living overseas to contribute to the city's material improvements. One of the responses obtained by the society was the donation of $100 and a letter from M. Camacho Roldán, in which he expressed his joy for the planting process undertaken by this society in the Camellón de San Victorino promenade and proposed to use *cajeto* trees (*Citharexylum subflavescens*) for its ornamentation. Camacho Roldán asserted that the tree was abundant in the neighbouring town of Ubaque across the Eastern Hills, and he highlighted that the large shadow produced by its treetop, its fast-growth rate, and its non-prejudicial roots would provide sufficient advantages for the *cajeto* to be seriously considered as a raw material for use in adorning the promenade.[62]

The recognition of native flora had begun to increase by the end of the 1910s; however, suggestions such as that expressed by Camacho Roldán or the SMO were scarce, and efforts to endow the city with local trees did not have a significant outcome. Despite the abundance of potential ornamental plants suitable to be acclimated in the urban environment of Bogotá, the lack of resources, the European influence and the pragmatic approach to the species used in the city's gardens, promenades, and parks did not promote the exploration of native local floristic richness. This contradiction between Bogotá's extremely rich flora and the historical influence of Western and temperate nature is revealed in the words of Genaro Valderrama. In 1886, seeking to replace Casiano Salcedo as the man in charge of the administration of the city's parks and public gardens, he wrote:

> I must remark that the cold weather areas in the country have a large variety of trees, much more beautiful indeed than the European ones. These [European trees] present, in general, a lugubrious and unpleasant appearance. I could mix them each other, creating with this a very nice outcome which is not presented in the European parks.[63]

As Valderrama's words indicate, the idea of a mixed nature was already present in the minds of some Bogotanians. However, two years later, when proposing a budget to undertake the improvements in the city's parks and gardens, besides stating that he would reduce monthly spending by nearly half, Valderrama observed:

> From that perspective, if the government accepts this proposal and decides to improve these parks and gardens bringing seeds, plants and

tress from Europe, I will provide a list of these plants soon. In doing so, I will be able to improve and rearrange (the parks and gardens) and conform to the rules of art.[64]

Despite attempts to follow Western modernist trends based on the use of European flowering plants and trees in the production of gardens and parks in Bogotá, local social, ecological and especially economic conditions required the inclusion of native species. In that context, although difficult to establish the number and use consistency, it is fundamental to recognise the importance of the native and naturalised flora as an unavoidable botanical presence in the city.

Notes

1 Mejía, *La Ciudad de Los Conquistadores 1536–1604*.
2 About the creation of a European vision of American nature, see Edgardo Pérez, *La Obra de Dios y El Trabajo Del Hombre: Percepción y Transformación de La Naturaleza En El Virreinato Del Nuevo Reino de Granada, Colección Bicentenario de Antioquia 32*, 1ra ed. (Medellín: Universidad Nacional de Colombia, 2011).
3 See Arturo Almandoz, *Planning Latin America's Capital Cities, 1850–1950, Planning, History, and the Environment Series* (London: Routledge, 2002).
4 An example of this aesthetic and material flow from Europe is represented in the Almanaque de Bogotá [Almanac of Bogotá] published by José Vergara and José Gaitan, see José Vergara and José Gaitan, *Almanaque de Bogotá, Guia de Forasteros Para 1867* (Bogotá: Imprenta de Gaitan, 1866).
5 Marshall Berman, *All That Is Solid Melts Into Air. The Experience of Modernity* (London: Verso, 1982).
6 Felix Driver and Luciana Martins, *Tropical Visions in an Age of Empire* (Chicago: University of Chicago Press, 2005).
7 Claudia Juranville, "Botánica, cartas de una directora de colejio a una antigua discípula," *La Escuela Normal, periódico oficial de instrucción pública (Bogotá)*, Febrero 3, 1872.
8 About cultural differences in approaches to nature, see for instance Yi-Fu Tuan, "Discrepancies between Environmental Attitudes and Behavior: Examples from Europe and China," in *Man, Space, and Environment. Concepts in Contemporary Human Geography* (London: Oxford University Press, 1972), 68–81.
9 About the concept and the implication of the Social Construction of Nature' concept, see David Demeritt, "What is the 'Social Construction of Nature'? A Typology and Sympathetic Critique," *Progress in Human Geography* 26, no. 6 (December 1, 2002): 767–90, https://doi.org/10.1191/0309132502ph402oa.
10 Charles Saffray, "Viaje Por La Nueva Granada," in *Colombia En Le Tour Du Monde (1858–1876)*, ed. Pablo Navas Sanz de Santamaría, vol. I (Bogotá: Villegas Editores, 2013), 162–63.
11 José Asunción Silva, *Poesía y Prosa: 'De Sobremesa'* (Bogotá: Atenea, 2015).
12 Luisa Garavito, ed., *Poesía Romántica Colombiana: Antología, En Homenaje a Los Grandes Poetas Románticos Colombianos, Colección Poesías Del Alma*, Primera edición (Bogotá: Multicopias Editores, 2012).

13 The recognition of tropical nature as a source of inspiration found its space in the poetry of Aurelio Arturo and Raúl Gómez Jattín, see Raúl Gómez Jattín and Carlos Monsiváis, *Amanecer En El Valle Del Sinú: Antología Poética, Tierra Firme* (Bogotá: Fondo de Cultura Económica, 2004); Aurelio Arturo, *Morada Al Sur, Un Libro Por Centavos (7)* (Bogotá: Universidad Externado de Colombia, 2004).

14 Mindrelio, "Un crítico impertinente," *La siesta (Bogotá)*, Abril 13, 1886.

15 José María Gutiérrez de Alba, *Cartilla Agraria ó Tratado Elemental de Agricultura y Ganadería. Dedicado a La Juventud Colombiana* (Bogotá: J.B Gaitán, 1878), 48.

16 Gutiérrez de Alba, *Cartilla Agraria ó Tratado Elemental de Agricultura y Ganadería*, 48.

17 Correspondence between Robert Thomson and Minister of Development, January 30, 1893, folio 25, tomo 823, fondo Ministerio de Obras Públicas, AGN.

18 Correspondence between Robert Thomson and Minister of Development, January 30, 1893, folio 27, tomo 823, fondo Ministerio de Obras Públicas, AGN.

19 See http://catalogoplantasdecolombia.unal.edu.co/en/resultados/familia/Ericaceae/.

20 Report of activities between 1892 and 1894 by Casiano Salcedo, June 1894, folio 23, tomo 823, Fondo Ministerio de Obras Públicas, AGN.

21 In order to classify the geographical origin of the plants used in the construction of green spaces in Bogotá, I use the geographical division of the introduced plant origins posed by Richard Brummitt. In order to establish the origin of plants he suggested eight global regions: (1) Europe, (2) Africa, (3) Asia Temperate, (4) Asia Tropical, (5) Australia, (6) Pacific, (7) Northern America and (8) Southern America, see Richard Brummitt, *World Geographical Scheme for Recording Plant Distributions*, 2nd ed. (York: TDWG, 2001).

22 Proposal to administrate the parks and city's gardens made by Agustín Torres. February 28, 1894, folio 58–9, tomo 823, fondo Ministerio de Obras Públicas, AGN.

23 Ruperto Ferreira, "Contrato número . . . De 1896 celebrado con el Sr. Casiano Salcedo sobre conservación y mejoras de parques y jardines públicos en la ciudad," *Diario Oficial (Bogotá)*, Julio 4, 1896.

24 Contract between Bernado Favera and Minister of Economic Affairs, July 25, 1896, folio 277–8, 302–3, tomo 823, fondo del Ministerio de Obras Públicas, AGN.

25 During the first decades of the twentieth century, in keeping with the consolidation of the Latin American nation-state after the Independence Centenary, plants turned into political elements, which entailed a complex resignification of the natural qualities. The conception of the national identity spread to encompass the national nature, and as a consequence, many plants and flowers were adopted as motherland symbols. For example, see Enrique Pérez Arbeláez, *Paisajes, Tierras y Trabajos* (Bogotá: Editorial Minerva, 1948), 199–201.

26 Also in the inventory of the gardening element of this park he mentioned how they only had two watering cans in very poor condition.

27 Report made by Cassian Salcedo when he gives the administration of the parks and public gardens to Bernardo Faverna, S.F., folio 306–312, tomo 823, fondo del Ministerio de Obras Públicas, AGN.

28 Correspondence between Bernardo Faverna and Minister of Economic affairs, August 12, 1896, folio 314–19, tomo 824, fondo Ministerio de Obras Públicas, AGN.
29 Report made by Genaro Valderrama to Minister of Economic affairs, September 30, 1896, folio 314, tomo 823, fondo Ministerio de Obras Públicas, AGN.
30 Correspondence between Bernando Faverna and Minister of Economic Affairs, April 2, 1897, folio 335–37, tomo 823, fondo del Ministerio de Obras Públicas, AGN.
31 In the archives it is not clear where the extra money used to cover the park administration deficit came from, see Final contract between Casiano Salcedo and Minister of Economic Affairs, April 8, 1897, folio 223, tomo 823, fondo del Ministerio de Obras Públicas, AGN.
32 Something is unresolved at this point: why they cancelled the contract with Faverna when he asked for an increase in the contract and afterwards the Ministerio de Hacienda signed a new one with Salcedo for more than the double prize?
33 Report made by Casiano Salcedo where he quite the contract to run the management of parks, November 8, 1897, folio 204–9, tomo 283, fondo del Ministerio de Obras Públicas, AGN.
34 About the role of the Commercial Houses as a link between Latin America and Europe, see Romero, *Latinoamérica Las Ciudades y Las Ideas*, 291.
35 "Arboles, Flores y Frutas (Rochester's Alive Plants Company Advertising)," *El aviso (Medellín)*, Junio 9, 1897.
36 Correspondence between Casiano Salcedo and Minister of Economic Affairs, July 10, 1897, folio 189, tomo 823, fondo del Ministerio de Obras Públicas, AGN.
37 Walter Benjamin, "Paris: Capital of the Nineteenth Century," *Perspecta* 12 (1969): 165–72; Octavio Paz, *El Laberinto de La Soledad, Postdata y Vuelta al Laberinto de La Soledad*, 3ra ed. (México, DF: Fondo de Cultura Económica, 1999); Romero, *Latinoamérica Las Ciudades y Las Ideas*.
38 "Otro," *Gaceta de Colombia (Bogotá)*, Febrero 21, 1830.
39 Correspondence between Byron Fish and the Minister of Economic Affairs, January 22, 1898, folio 81, tomo 823, fondo del Ministerio de Obras Públicas, AGN.
40 Resignation letter made by Casiano Salcedo, October 16, 1897, folio 203, tomo 823, fondo de Ministerio de Obras Públicas, AGN.
41 Ibid.
42 Legal complaint made by Augusto Torres on behalf of Casiano Salcedo, April 1899, folio 218–19, tomo 823, fondo del Ministerio de Obras Públicas, AGN.
43 Contract between José Leocadio Camacho and Gabriel Peña Solano, November 28, 1897, folio 89, tomo 823, fondo de Ministerio de Obras Públicas, AGN.
44 Correspondence between Julio Garavito and Minister of Economic Affairs, March 6, 1901, folio 381, tomo 823, fondo del Ministerio de Obras Públicas, AGN.
45 Eduardo Posada Carbó, ed., *Colombia. La Apertura al Mundo, América Latina En La Historia Contemporánea*, vol. III (Barcelona: Fundación MAPFRE & Taurus, 2015), 20.
46 Some of the publications made by Izquierdo are Izquierdo, *Memorial Sobre Agricultura*; Izquierdo, *Riqueza Nacional: El Caucho*; Antonio Izquierdo, *Estudio Sobre Bosques* (Bogotá: Linotipos de 'El Diario Nacional,' 1918).

47 Vegas y Ávila, "Reportaje con el Sr. D. Antonio Izquierdo I," *El Correo Nacional (Bogotá)*, Mayo 14, 1908.
48 Vesga y Ávila, "Reportaje con el Sr. D. Antonio Izquierdo II," *El Correo Nacional*, Mayo 15, 1908.
49 Vesga y Ávila, "Reportaje con el Sr. D. Antonio Izquierdo IV," *El Correo Nacional*, Mayo 18, 1908.
50 There is an interesting fact about the *hevea* rubber tree sent by Izquierdo to Colombia. He mentions how, along these the seedling, he sent dry vouchers of these plants in order to be used as an example followed by the Colombian people to search for this kind of plant in the mountains. However, the rubber plant is a tree naturally growing in the Amazonian region of Colombia. Results were paradoxical that native tree seedlings were sent from Ceylon, where the British had established large plantations from seed coming from the Amazon. The same can be said about the cacao beans sent equally from Ceylon or the potatoes imported by Izquierdo from England.
51 Vesga y Ávila, "Reportaje con el Sr. D. Antonio Izquierdo VI," *El Correo Nacional*, Mayo 26, 1908.
52 Izquierdo, *Estudio Sobre Bosques*.
53 The table is made with the original information given by Izquierdo. However, some places mentioned by the author present significant inconsistencies in the addresses as well as some misspellings.
54 Acta SMO, March 13, 1918, folio without number, libro 2, Actas MOB, MdC.
55 Charles Watkins, *Trees, Woods and Forests: A Social and Cultural History*, 1st ed. (London: Reaktion Books, 2014).
56 Acta Sociedad de Mejoras y Ornato [SMO], April 26, 1867, folio 6, tomo 1, Actas MOB, Archivo Museo del Chicó, Bogotá, Colombia [MdC hereinafter].
57 David Harvey, "Between Space and Time: Reflections on the Geographical Imagination1," *Annals of the Association of American Geographers* 80, no. 3 (September 1, 1990): 418–34, https://doi.org/10.1111/j.1467-8306.1990.tb00305.x.
58 Alejandro de Humboldt, *Viaje a Las Regiones Equinocciales Del Nuevo Continente Hecho En 1799, 1800, 1801, 1802, 1803 y 1804, Biblioteca Venezolana de Cultura*, vol. II (Caracas: Ministerio Nacional de Cultura, 1941), 31.
59 Actas SMO, December 11, 1918, folio 82, libro 3, Actas MOB, MdC.
60 Actas SMO, March 20, 1918, folio sin numerar, libro 2, Actas MOB, MdC.
61 Actas SMO, December 11, 1918, folio 83, libro 3, Actas MOB, MdC.
62 "Sociedad de Embellecimiento," *El Heraldo (Bogotá)*, Febrero 11, 1899.
63 Correspondence between Genaro Valderrama and the Minister of Infrastructure, Braulio Velez. April 16, 1886, folio 5, tomo 823, Fondo Ministerio de Obras Públicas, AGN.
64 Budget for the material improvements of parks and gardens made by Genaro Valderrama, April 25, 1889, folio 6, tomo 823, Fondo Ministerio de Obras Públicas, AGN.

7 New and disciplined relationships with plants

Introduced and native plants geometrically organised by self-taught gardeners endowed the city with a unique assemblage of plants hitherto unseen in Bogotá. But more than hygienic and ornamental devices, these syncretic flora in the guise of parks and public gardens became the scenario where new public relationships with plants unfolded. Thus, unlike previous interactions with plants in places such as patios and solares, in the late nineteenth-century Bogotá plants were planted first and, later, a regulated ensemble of suitable human–plant interactions were superimposed on it. This chapter explores how in modern Bogotá, plants became political actors with an important role in the discipline and control of human bodies that were forced to engage in a new set of ethno-botanical interactions.

The fenced nature

The physical characteristics of modern spaces define their potential uses by citizens. In contrast to the formerly plain squares, modern gardens required constant maintenance and care, which is one reason they were usually enclosed by iron fences. These structures, common in most nineteenth-century parks and public gardens, can be understood as manifestations of modern innovations. Beyond the clear expression of an emergent relationship with plants, parks and public gardens in modern Bogotá provided an opportunity to display the innovations produced as a result of industrial developments. For instance, electricity and its power to light the night was first revealed to the public in parks. Equally, since modernity has a lot to do with materials and the power to cast them, parks were places to exhibit goods, structures and machines made of iron and steel. Poles, kiosks, exquisitely carved fountains and also elaborated fences made part of the urban facilities presented in the nineteenth-century parks opened in Bogotá.

The presence of iron fences and other modern innovations in parks and public gardens was a symbol of physical improvement; thus, a fancy iron fence replaced the barbed wire previously used for the protection of gardens

DOI: 10.4324/9781003208167-10

in the Plaza de Bolívar (see Figure 4.5). However, even as iron fences protected plants against free wandering animals such as cows, they also served as a symbol of exclusion and spatial control. Just as the *boulevards* of Paris were partially created as a means to deter the barricades,[1] Bogotá's public squares were turned into gardens that limited people's access and avoided the congregation of citizens' bodies as a manifestation of their social and political expression. In this regard, Fabio Zambrano explains:

> The modernisation process whereby the old squares were transformed into parks was not accompanied by political modernity. As a consequence, in the public spaces, the enclosed prevailed upon the characteristic aperture of the early republican city. . . . In the (late) republican squares, aesthetic pleasure contemplation is introduced. Turning them into gardens with fences and controlled access would prevent those spaces from being used as a gathering scenario for the political population.[2]

The replacement of open public spaces (namely the old colonial squares) with enclosed public spaces (namely modern gardens), meant not only the political control of bodies but also a disruption of a myriad of social relationships and daily rituals and activities. As described in the preceding section, this alteration of urban rhythms was first evidenced in the displacement of Bogotá's markets. Like the case of *Les Halles* in Paris,[3] markets were targeted by a medical guild immersed in the environmental notion of health and disease, which denounced such settings as sources of moral and physical decay and thus ensured their relocation to roofed and regulated spaces. However, the outdoor markets served a purpose that extended far beyond their mercantile functions; they were sites of spontaneous gatherings attuned to the citizens' emotional and political feelings (i.e. Lefebvre's 'living spaces'). With the appearance of fenced gardens, these dynamics drastically changed, concomitantly transforming the citizens' 'sense of place' by forever reshaping their particular and personal ways of living and understanding certain urban spaces and the plants growing on them.[4]

Parks and public gardens as alternative and restricted modern amenity

Whereas many European and even some Latin American capitals had developed robust cultures of leisure and entertainment by the second-half nineteenth century, Bogotá was an isolated, convent-like, and—at least in the eyes of travellers—boring city.[5] The latter is highlighted again and again in the writings of foreigners who complained about the lack of 'things to do' in the capital. As Jorge Brisson, the French engineer hired by the Colombian government, affirmed in the 1890s: 'The distractions are null, at least for the foreigner without social relationships that prevent him from joining the social gathers within some houses. At the moment of my arrival, a bad Spanish Zarzuela company left the city . . .'[6]

Besides Catholic celebrations, few social activities gathered large groups of people together. Among the most significant social encounters were the so-called *tertulias*, social gatherings occurring in private homes and explicitly devoted to discussing social life, politics, and art. The *tertulia* was a well-developed practice among the upper class and had also been adopted by the newly emerged artisans' guild. From time to time, some of Bogotá's wealthiest families also organised parties known as *bailes* (balls) to celebrate weddings, baptisms, or other family events, often in a manner that disrupted daily life in the small city. *Bailes* were used as a space of exhibition for elites,[7] as was the city's most reputed social spectacle: the opera.

The main theatre of the city, *Coliseo de Bogotá*, was built by a private investor at the end of the eighteenth century and began hosting several itinerant opera companies in 1858.[8] It was in the candle-illuminated *Coliseo* where the city's elites displayed their dresses and manners advertising their social status. Although the opera in Bogotá did not reach the levels of public spectacle exhibited in other Latin American capitals such as La Habana or Mexico City, it was nonetheless considered as a 'civilising spectacle' by the new urban elite.[9] Such forms of social amusement had been consumed by and devoted to the city's wealthy since colonial times, and members of the lower classes were excluded from such events. The cultural marginalisation of non-elites was eloquently depicted by José María Cordovez, among the most colourful of the city's nineteenth-century chroniclers:

> Nowadays only two kinds of people attend the theatre, the wealthy and privileged and those that pretend to be rich (who knows how). The middle-class families and the artisans cannot spend what they earn in several days of work to get the enjoyable and instructive pleasure to attend, at least once per month, to this sort of amusement because of the high price of the tickets.[10]

Cordovez also mentioned the social benefit that opening the opera performances to poor people would have allowed:

> We should say that the entrepreneurs had ignored the advantage for them and for good manners if they would set up a theatre section for the labouring people. This access would foster among them the pleasure for this kind of entertainment, drive them away from the taverns, places frequented by them because of the lack of honest and affordable distractions.[11]

Whereas fine arts such as the opera were almost exclusively consumed by the elites, one of the few options left for the newborn working class was the pub. Known in Bogotá as *chicherias*, in addition to being the artisanal factories where *chicha* (indigenous-origin fermented alcohol from maize) was produced, these pubs served as gathering points for the popular classes where the rigid patterns of social life imposed by Catholicism were relaxed. As a consequence, popular music, sex, fights, and political discussions flourished in the *chicherias*, turning them into 'pagan sub-worlds'.[12] Furthermore, commonly produced under dubious hygiene conditions, the *chicha*s' alcohol levels were uncontrolled, and excessive drinking often led to constant quarrels and street fighting. Moreover, the surroundings of the *chicherias* were used as latrines due to the lack of sanitary facilities. Thus, in the eyes of doctors and politicians, the *chicha* (and by extension, the *chicherias*) were among the most notorious causes of social and 'racial degeneration'. Consequently, many campaigns were established in order to eradicate this vice from the city, although they were almost always very inefficient. Among historically documented campaigns were efforts to persuade the popular classes to abandon the consumption of *chicha* for 'healthier' beer, which was even recommended for pregnant women.[13]

Parks and public gardens were identified as spaces with the potential to promote 'popular education' for the labour class and keep them away from the *chicherias*. For example, in 1918, after repeated failures in the public administration of the parks and public gardens, the local government decided to transfer the administration of these spaces to the Sociedad de Mejoras y Ornato. Due to continued economic constraints, the earliest meetings held by the SMO featured debates regarding the establishment of an entrance fee for the city's parks to help in their maintenance. This proposal received sharp criticism, particularly from Mr Araujo, who stated that 'It is very risky the idea of authorising the Mistress to collect money for the park's entrance. . . . 'To charge for entrance would mean to close them to the poorest; which then would end up at the taverns and this society must, it is its inevitable obligation, to avoid this at any cost.'[14]

Initially, parks and public gardens were thought of as healthy places open to all of the population; the oxygenated air produced by the plants would work to produce healthy people. However, the opening of the parks to all people was only partly true. Parks and public gardens posed a decisive social challenge in a manner different from other modern innovations. For instance, whereas the upper class was the first to enjoy the water supply and the iron pipe sewer system, they had to share 'the privilege' of using parks with the rest of the population (see Figure 7.1). Consequently, parks and public gardens were the first modern place where different social classes

Figure 7.1 Anonymous, *Parque de Bolívar*, ca. 1890, postcard.
Source: Editores Duffo

could meet and mix with each other. At least from a hypothetical perspective, the gardens were there to be enjoyed by the housemistresses who have had a long tradition of gardening practices in their patios, as well as for the people whose unique relationships with plants had been through the extraction of wood, charcoal fibres, and other products from the nearby forest. However, such inter-class exposure in a rigidly hierarchical society such as that of Bogotá entailed significant conflicts. A set of restrictions on the use of parks was established in order to mediate the social contestation that emerged within the modern public space, which manifested in efforts to maintain both their social convivence and physical condition.

The establishment of urban parks meant the creation of a new bureaucratic body responsible for their administration and the people's behaviours regarding others and plants. This new administrative office was formed by the Ministry of Infrastructure, an administrator and an intervenor (park inspector). Additionally, this office relied on the work of several gardeners, some *celadores* (guards) and police officers. The diversity of people involved in the care and administration of parks reflected their status as places where the care and maintenance of plants intermingled with the care and vigilance of the human bodies that frequented them. Therefore, whereas gardeners took care of plants, *celadores* and police officers were in charge of bodies; it was their responsibility to watch over the social behaviour at and general social usage of parks. Given the large range of responsibilities entrusted to the *celadores*, they played a key role in the functioning of parks and public gardens. For example, they had to open the parks' gates at 6 am and close them at 7 pm, as well as 'correct with moderation and good and civil manners, the offences that guided by ignorance some of the parks attendants commit. In case these offences were serious he must inform the park inspector and the authorities to its efficient correction'.[15] In addition to these duties, the *celadores* had to undertake daily inspections of the park's facilities, such as fences, poles, and paths, and they were required to supervise the planting, pruning and cleaning process made by gardeners.[16] As a consequence of this workload, plus their reduced numbers, the available *celadores* in Bogotá were insufficient to control an 'uneducated population' that destroyed the gardens and constantly misbehaved.[17]

The lack of social control in the parks and public gardens led to a constant need for police officers to oversee the 'immoral behaviours' of certain visitors. Thus, the police presence was very important for the parks' administrators. For instance, when the SEB was in charge of the parks, they decided to close most of them in 1899 due to the lack of police officers. In one letter sent to the Ministry of Developmnet, the director of the Society, José Pérez, wrote:

> Answering your past 28th of June notification, I have to inform you that the board has decided not to order the opening of parks because of the lack of police to custody them. This decision has been taken since, without this requirement, they will be destroyed by the public.[18]

The police department was suffering the same economic challenges as most of the government departments, and the police presence in the parks was rather sporadic. The police's frustration with this situation is revealed in another letter sent by the Police director to the Minister of Development in 1894:

> It is too hard to take care of the Parque Centenario, because of its large extension, and, also, because it is full of trees, destroyers can hide themselves easily. Once the *celadores* go home at the end of the day, they must leave the keys in the closest police station in order for officers to be able to surveil them from time to time. It is a real problem to undertake a proper vigilance of those places since the number of police officers devoted to this service are few.[19]

What kinds of misbehaviours were carried out within parks to make the police officers' presence so important? One of the main problems faced by the park's administrator was the destruction of lawns due to the constant transit of people outside the established paths. In addition, the *celadores* complained about rubbish left by some visitors and the constant thefts that occurred in the parks. With respect to the latter, some people tended to uproot flowering plants to take them home during the day, and at night, people often climbed the iron fences and stole some of the very few gardening tools and other elements from the park facilities. However, the physical destruction of parks was not only perpetrated by adults and thieves. In the early twentieth century, the vigilant of the Parque de la Independencia argued:

> I have the honour to inform you that the number of police officers that are assigned to this park daily is completely insufficient given the large assistance in the recent days. Even more, if you consider that eighty percent of kids come to produce damages in the park, mistreating trees, throwing stones at the ponds and destroying the lawns.[20]

Beyond the material damage produced by thieves or unruly children, one of the most common sources of concern for the park administrators and the police was offences against morality: people urinating against the trees, having sexual assignations at night when parks were closed, drunks fighting, and recurrent catcalling were behaviours that did not fit into places thought of as symbols of civilisation. In this regard, some of the *Señoras of Bogotá* (ladies of Bogotá) publicly complained in 1890 that 'We will have to cancel our hygienic walk in the Parque Santander since the students visiting the park do not behave in an expected way'.[21] Although the ladies did not specify what kind of behaviour took them to cancel their 'hygienic walks', three years later, some *padres de familia* (family men) wrote a letter protesting because:

> The medical students that attend the Parque de Los Mártires during the morning read aloud and with clear intentionality the anatomic chapters addressing the most hidden parts of the human body, harming with this the shyness of the Misses that visit the park.[22]

However, the medical students did not precisely belong to the lower class, and their actions were not viewed as being equally problematic as those undertaken by the popular and socially excluded classes during their visits to the parks. For instance, in 1895, scandalised by the moral affronts tolerated by the elegant people of the city during their visits to the parks, Genaro Valderrama, the General Inspector of Parks and Public gardens commented:

> I have to manifest to Your Honour that in some parks and squares concur convene beggars and lepers to use the benches located there without any

> possibility to forbid them the entrance given the freedom they have to do so. This must be solved through the expedition of tickets used to control the entrance. The ticket, supplied by a high hierarchy employee, would also avoid the depravations committed there by uncultured people.[23]

Four years later, Valderrama's views had become even more radical. In 1899, he went beyond regulation through the use of tickets to propose entrance restrictions as a means to deter undesirable people from the Parque Santander:

> The number of beggars and lepers that visit the park offer a repugnant view. This prevents decent people to sit on the benches. For this reason, the entrance to these places to these kinds of people, and, furthermore, this restriction must apply to *los de ruana*, who only attend the parks to steal what they can and to destroy the benches to take the planks.[24]

Los de ruana was the name given to the poor and excluded people who maintained their traditional way of dressing with *ruana*, a traditional Andean poncho-style outer garment (see Figure 7.2).[25] The general description of *los de ruana* encompassed a set of characteristics that applied to the vast majority of the city's population. From the elite's point of view, they were *chicha* addicts, illiterates and thieves lacking any kind of civic education; in summary, they represented everything but modernity. Moreover, *los de ruana* represented the opposite of the civilised people known in Bogotá as *cachacos*. The denomination as a *cachaco* was commonly used to designate a high and middle-class male with some kind of formal education, who, as a general rule, had adopted the way of dressing and the manners of the leading European societies (see Figure 7.2).[26]

Despite repeated attempts to exclude *los de ruana*, there is no real evidence suggesting the existence of any systematic mechanism of exclusion in the parks. Nevertheless, the opinions expressed by Valderrama reveal the profound contradictions embedded in Bogotanian modernity. For example, throughout his public life, Valderrama defended the use of plants and trees as elements of a moral and physical fortification based on the medical thinking of the time; however, it seems that people had to first be turned into modern citizens to be able to enjoy the beneficial airs produced by the parks' ornamental plants and trees. For Valderrama, *los de ruana* could be morally and physically strengthened by the influence of urban nature; in other words, they could be cleaned and civilised through their exposure to the good airs produced by plants. However, the contradiction represented by Valderrama seems most stark when considering the ill people who visit parks. Although parks were regarded as a suitable place for the

Figure 7.2 Anonymous, *Los de ruana* (left) and *cachacos* (right), 1918, photograph.
Source: Cromos Magazine 1918. Courtesy of the Biblioteca Nacional de Colombia

ladies' 'hygienic walks', when lepers used those spaces to rest, the directionality of the influence was inverted. Now, the influence of the vegetation that purified and embellished those spaces was undermined by the lepers, who made the parks ugly and undesirable places. However, as evinced in the next section, once the lepers and other kinds of marginalised people had been enclosed in suitable institutions, the beneficial actions of nature were restored.

The attempts to promote a disciplined use of green spaces were accompanied by educative campaigns tending to create a new attitude toward plants. Using private organisations such as Bogotá's Embellishment Society, local elites orchestrated large campaigns of green proselytism. Focused primarily on the importance of taking care of trees, this campaign crystallised in the so-called *Fiesta del* Árbol (Tree celebration), which took place for the first time in 1912. This event was scheduled as a section of the Independence Celebration acts on the 20th of July and was particularly devoted to children.[27] In the celebration, children undertook a massive plantation of trees in urban parks, after which they sang the national anthem and other songs exalting the symbolic link between trees and children: like children, the recently planted trees were now part of the nation; and like trees, children would grow yielding fruits. A 1912 publication that covered the first Fiesta del Arbol published an image of this day with the following caption

> Aspect presented by the Bosque de la Independencia during the Fiesta del Árbol. Here children from Bogotá's schools sang the national anthem while standing in front of the Simon Bolívar statue. More than three thousand students and numerous spectators make part of this celebration (see Figure 7.3)

The celebration of the tree party specially designed for children shows how plants, but specially trees turned into pedagogic tools that were constantly used within the classrooms as a way to present creating desirable

Figure 7.3 Children gathered around the Simón Bolívar statue as part of the *Fiesta del Árbol* Celebrations, 1912, photograph.

Source: El Gráfico no. 94, Julio 20, 1912. Courtesy of the Biblioteca Nacional de Colombia values. Poems and catechism-like tree prayers were usually impressed and published in local magazines to be taught to scholars. In sum, part of the green campaign was to link the trees with the national feelings and with the marked catholic culture present in Colombia.

The vegetal proselytist aim reached its highest level during the Fiesta del Arbol in 1912; however, the role of trees as a coadjutant in the education of the young can be tracked decades before. For instance, in 1881, the Ministro de Fomento y Educación (Minister of Education and Development) ordered each of the public primary schools across all the country's territories to undertake a large planting campaign of the recently introduced eucalyptus trees (see next section). To support this mandate, the Ministry distributed a publication called *El Eucalipto*. Written by the Algerian citizen Juan Carlos Manó, this publication sheds light on general ideas of the cleaning power of eucalyptus as they prevailed at the end of the nineteenth century. In his report, Manó explained how the French colony in Algeria could thrive after planting eucalyptus trees as a way to fight malaria, and how following this example, Spain, Italy and Austria achieved excellent results in creating healthy spaces. Manó closed with the assertion that 'The landlord of places infested with fevers who does not plant eucalyptus trees commits the same crime against humanity (and against the homeland) as the father who does not immunise his child during a smallpox outbreak'.[28]

Manó's work as a governmental botany consultant was denounced in a bitter and justified critique by the Sociedad de Medicina y Ciencias Naturales (Medicine and Natural Science Society).[29] Nonetheless, his publication on eucalyptus persuaded many school principals of the benefits of planting these trees in their *patios*. Schools in Bogotá and in many other regions embraced the use of this tree that can contribute to the sanitation of spaces and therefore help in the development of children. One of the first to answer the Ministry's call was one of Manuel Ezequiel Gonzales, principal of the Colegio Mayor de Nuestra Señora del Rosario, the oldest and most prestigious school in Bogotá. Gonzales affirmed his promise with the successful establishment of eucalyptus trees at his school. Thus, the regulation concerning eucalyptus in education centres was successfully implemented, not only in schools but also in other enclosed places where the social improvement of people was pursued through the use of plants.

The role of plants in institutional spaces of reclusion

In modern Bogotá, the plants previously cultivated in patios and solares played roles that extended beyond their placement in the construction of public green spaces. Plants were viewed as means of physical and moral fortification; however, they also served as devices of discipline and social control. This means that garden production in Bogotá was not exclusively limited to public spaces such as parks. Rather, modern ways to perceive, use and perform with plants were also adopted as part of the consolidation of social institutions such as hospitals, prisons, and asylums, where plants acquired a clear political dimension when used for the regeneration and social reintegration of certain types of excluded people.[30]

The exclusion of certain groups of people from social life was not a new practice in Bogotá. Based on royal laws from the sixteenth century, the post-independence Republican government criminalised people who lacked any recognised skills, incomes or social positions, and who were labelled as vagabonds. Several bills were approved in order to tackle this 'problem', which had become widespread following the war for independence. The resulting punitive measures against 'idle' people included banishment, forced military service and incarceration. Additionally, in many cases, vagabonds and wandering people were also used as a labour force in the construction of the new urban infrastructure.[31]

Individuals afflicted with leprosy, goitres, elephantiasis or sexually transmitted diseases were similarly excluded from society. In the early years after independence, the new Creole ruling class sought to solve the problem posed by incurables by building hospitals located away from the main cities; however, this measure proved untenable due to post-war economic constraints. As a result, it was common to see lepers wandering around cities, where their entrance was often forbidden by physicians.[32] After the second half of the nineteenth century, lepers were finally confined in a hospital located 78 miles away from Bogotá, thus partly resolving one of the most disruptive problems for the urban population.

Like those with physical disorders, *lunatics* were also identified as a source of social disruption, and *melancholic, insensate, or insane* people were generally incarcerated in asylums established specifically for their confinement. However, the conditions inside these institutions were widely criticised and some of their practices against the inmates were considered to be forms of torture. For example, an 1835 account reported that the mentally ill were 'enclosed in cold and gloomy jails tied with chains, bathed by force with chilly water, in one word, tortured'.[33]

Those confined in jails or prisons experienced similar conditions. During the first republican decades, many Colonial buildings were used as city jails, the most famous of which was the San Buenaventura building. To replace this rammed-earth-wall structure, the government of Cundinamarca (the state encompassing Bogotá and six other governmental departments) undertook the construction of a modern prison in 1878. Commonly known as the *Panóptico*, this building made of bare rock housed a large number of political prisoners from defeated factions of the civil war even before its completion. Following their release, some former inmates, mostly from the Liberal Party, produced harrowing accounts deploring the health and conditions experienced by those incarcerated there. One remarkable example is the text *Los secretos del Panóptico* (The Panoptic Secrets) produced by Leon Gómez in 1900, in which he described 'The multitude of people, the small size of the place, the lack of air and water. Also, the general slovenliness, the absolute lack of sun and exercise had to develop diseases. Dysentery, smallpox and typhus had there their place'.[34] Many inmates died in prison due to the *Panóptico's* minimal provisions to cover basic human needs.

During a large part of the nineteenth century, those *others* who did not align with normalised social behaviours or social and physical conditions and biological characteristics were viewed as threats by the medical and political spheres. Consequently, they were excluded from society and forced to live in precarious conditions in institutions specifically designed to confine them. However, after the consolidation of the country's medical schools and the importation of determinist ideas concerning the environment, the government sought the transformation of bodies through their confinement in oxygenated and beautiful spaces. Ill people, orphans, and criminals, for instance, could be morally and physically cured through their contact with a calm and country-like environment. Therefore, nature in general, and plants, in particular, were used as therapeutic elements that by rehabilitating people also contributed to the social progress of the city.

In accordance with the environmental ideas presented in the first part of this chapter, the medical guild was responsible for identifying the most convenient spaces to establish asylums, hospitals, orphanages, and jails. Although doctors considered many environmental characteristics when determining suitable locations, the most important of these were related to the purity of air and wind currents. For instance, when a group of doctors was consulted about the most suitable place to open a new mental asylum in Bogotá in 1884, their report suggested moving the old mental asylum towards the north side of the city close to the Parque Centenario. As they argued,

> the San Diego place is the most visited *paseo* (promenade) in the city. There, people from all the backgrounds attend for exercising, families send their kids to breathe clean air, and the council is planting a beautiful park, the Centenario, which would be one of the best ornaments of the city.[35]

In addition to the location of the confinement institutions, building design and general structure were also of great importance. Hospitals, orphanages, and asylums had to be ventilated, illuminated by sunshine, and endowed with an efficient drinking water system and a sewer. The presence of plants was also a critical element. In the aforementioned report concerning the new mental asylum in Bogotá, doctors highlighted that '[t]he buildings must be separated by planting trees on the streets.'[36] Therefore, in order to achieve the best results in terms of healing, hospital infrastructures must be endowed with trees.

The role of plants as an integral part of the therapeutic treatment applied to those confined people continued into the twentieth century. In 1918, the Asamblea Departamental (District Government) named a commission to judge the performance of the government-supported charity institutions in Bogotá, which oversaw the care of orphans and the physically and mentally ill. In a section of the report describing the physical conditions of the Hospital San Juan de Dios, they highlighted:

> We visited the labyrinth-like hospital for three hours. Under any circumstance the hospital has the minimum conditions required by a place of its class that must be built conforming to modern innovations and in an adequate place. Everything there is asking to be translocated to another place. The hospital lacks gardens and groves. It lacks a large enough field where the ill can get better with the free and pure air and the view of vegetation. . . . A hospital without gardens and close lawns *where nature works in the strengthening of those hospitalised* will never be a model for these kinds of institutions, although many say the opposite [emphasis added]![37]

The commission also paid special attention to the improvements underwent in the Las Mercedes mental hospital, emphasising the verifiable improvement presented by the people confined there, which they mainly attributed to the adjacent plot cultivated with flowers. Furthermore, when presented with reiterative petitions from the hospital's neighbours asking to turn the plot into a public park, the commission suggested denying these requests because:

> The plot we were talking about is a terrain cultivated today by those miserables that lost the spiritual calm with the loss of their mental faculties. This calm only returns to them momentarily with the after-eating walks. The freshness of the plants cultivated there and the flowers with their essences make softer the violent vigour and the desperation of the lunatics. We say it again, without fields and flowers enjoyable to the sight of the ill, it is a cruelty to establish hospitals and, moreover, lunatic asylums.[38]

The therapeutic role of plants was not only defined by their ability to clean spaces and therefore bodies. Through agriculture and horticulture practices, plants played a significant role as corrective instruments. Growing vegetables and tending flowering plants were viewed as a means to cultivate discipline and a work ethic. For this reason, plants were an integral part of penitentiary and reformatory institutions. This relation between moral correction and plants is well exemplified by the multiple attempts to get rid of the city's beggars and vagabonds. In 1840, for example, masses of the unemployed and 'non-useful' were forcibly devoted to the cultivation of sugar cane and coffee in isolated spots.[39] In furtherance of this policy, agricultural penitentiary colonies were opened in vast isolated regions such as the Acacias Colony in the Orinoco Basin, where many criminals captured in Bogotá were used as a labour force in the cultivation of crops. However, not only criminals were suitable to undertake agricultural tasks. In accordance with ideas that related physical decay with moral weakness, agricultural practices were also perceived by doctors as a means to relieve the suffering and pains of the sick and incurable. In 1828, for instance, the well-recognised physician José Merizalde suggested the construction of lazarettos in different regions of the country where the sick could devote their days to agricultural work.[40]

Agricultural and horticultural practices were also applied as a means to shape orphans' moral and physical qualities. In the aforementioned report considering the charity hospitals' infrastructure of Bogotá in 1884, the physicians' board asserted the following regarding the Hospicio de Niños Desamparados (Orphanage of Abandoned Children):

> The institution lacks means to make of all those escaped-from-death children useful men. Any orphanage, to be complete, in order to be a real charity institution, must have arts and crafts studios. But, overall, they need to have spacious growing fields to transform the kids into healthy farmers, that once [they] go out of the cloister that hosted them, devote themselves to agricultural works.[41]

The Hospicio de Niños Desamparados also known as San José Orphanage was created by the Cundinamarca Charity Commission in 1882. The orphanage first used a location adapted for it in the village of Chapinero, which had provoked the above complaint about its precarious facilities. However, the Bill 39 of April 29, 1913, ordered the construction of a new orphanage. Construction began in 1917 on Bogotá outskirts in Chapinero; however, progress was delayed due to recurrent economic constraints and problems around the water supply, and the new institution was not opened until 1919. Given the possibilities offered by the new location, there was an increase in the activities seeking to make useful citizens of the orphans. As in the case of criminals and ill people, agricultural work was considered central to their social re-shaping process. The pasturelands surrounding the main building were transformed into a vegetable field where children grew vegetables. Additionally, in order to instil a sense of responsibility amongst the orphans, each was in charge of a particular furrow in which they had to grow and harvest their own vegetables (see Figure 7.4). As a *Cromos* magazine journalist described,

> Beyond the court, where kids had their crops, the pasturelands stretched. Each of those children was in charge of a fallow that must

Figure 7.4 Anonymous, *Asilo de Huerfanos de San José* [crop fields in the San José Orphanage], ca. 1920, photograph in glass, 8 × 10 cm.

Source: Courtesy of the Biblioteca Luis Ángel Arango, Bogotá

> be cultivated under the Director's surveillance. This sought to stimulate in those future-less kids the love for the countryside, the earth and the agriculture.[42]

Whereas agricultural activities were common in other institutions of this type, the San José Orphanage is remarkable due to the outstanding levels of horticulture and gardening reached within this charity institution (see Figure 7.5). Although most of the evidence remains only in the form of photographic traces, it seems that in addition to being responsible for the gardening of the ancient colonial squares, the self-taught gardener Casiano Salcedo also oversaw the San José Orphanage's horticultural works. This could explain, to some extent, the uncommon level of sophistication presented in the orphanage's gardens, manifested, for example, in the presence of topiary and the utilisation of climbing vines to cover the building walls, a practice commonly used in the construction of British gardens.

The remaining visual traces of the San Jose Orphanage gardens reveal how the creation of green spaces through the manipulation of plants was

Figure 7.5 Anonymous, *Asilo de Huérfanos de San José* [Gardens of the San José Orphanage], ca. 1920, photograph in glass, 8 × 10 cm.

Source: Courtesy of the Biblioteca Luis Ángel Arango, Bogotá present not only in Bogotá's parks and public gardens but also in cloister-like institutions devoted to the improvement of some social actors. In sum, the establishment of trees in mental asylums and gardens in orphanages elucidates the intimate relationship between bodies and space through the use of plants as key linkages. Also, parodically, the use of plants in the creation of good citizens helped to maintain the city agriculture practices that were in the process of being eradicated from the city. Consequently, these small kitchen gardens helped to maintain some of the botanical diversity that by that time was disappearing with the urbanisation of solares and peri-urban areas.

Notes

1 Walter Benjamin, "Paris: Capital of the Nineteenth Century," *Perspecta* 12 (1969): 165–72.
2 Fabio Zambrano, *Construcción Del Espacio Público Tres Parques de Bogotá: Nacional, El Tunal, Simón Bolívar* (Bogotá: Alcaldía Mayor, 2003), 49.
3 Green, *The Spectacle of Nature*, 47.
4 About the construction of *the place,* see Tim Cresswell, *Place: An Introduction*, 2nd ed. (Chichester, West Sussex: John Wiley & Sons, 2015); Yi-Fu Tuan, *Space and Place. The Perspective of Experience* (Minneapolis and London: University of Minnesota Press, 1977).
5 The conception of Bogotá as a boring city described by several travellers ignores popular class entertainment which includes the bull and roster fighting accompanied by a significant consumption of chicha (alcoholic beverage obtained after the fermentation of maize), see Jorge Humberto Ruiz, *Las desesperantes horas de ocio: tiempo y diversión en Bogotá (1849–1900), Opera eximia*, Primera edición (Bogotá, DC: Editorial Pontificia Universidad Javeriana, 2021).
6 Jorge Brisson, *Viajes Por Colombia En Los Años 1891 a 1897* (Bogotá: Imprenta Nacional, 1899), 157.
7 Cordovez, *Reminiscencias Escogidas de Santafé y Bogotá [Recurso Electrónico]*, 15–40.
8 Rondy Torres, "Tras las huellas armoniosas de una compañía lírica: La Rossi-D'Achiardi en Bogotá," *Revista del Instituto de Investigación Musicológica "Carlos Vega"* 26, no. 26 (2012): 162–200.
9 Ibid.
10 Cordovez, *Reminiscencias Escogidas de Santafé y Bogotá [Recurso Electrónico]*, 15–40.
11 Ibid.
12 Beatriz Castro, *Vida pública en las ciudades republicanas* (Bogotá: Grupo Editorial Norma, 1996), 259.
13 Ibid.
14 Actas SMO, March 20, 1918, Sin folio, libro 2 (1918–19), Actas MOB, MdC.
15 Regulations for Parks, Public Gardens and Promenades of the City, Abril 1882, folio 123, tomo 857, fondo del Ministerio de Obras Públicas, AGN.
16 Regulations for Parks, Public Gardens and Promenades of the City, Abril 1882, folio 123, tomo 857, fondo del Ministerio de Obras Públicas, AGN.

17 The conflict between control and disobedience in the parks reflects how under certain circumstances, the discipline of the bodies presented by Michel Foucault is surpassed by the disobedience of the everyday life attested by Michel de Certeau. Michel Foucault, *The Birth of Biopolitics: Lectures at the Collège de France, 1978–1979: Lectures at the College De France, 1978–1979*, ed. Arnold I. Davidson, trans. Graham Burchell, 2008 ed. (New York: Palgrave Macmillan, 2010); Michel de Certeau, *The Practice of Everyday Life*, trans. Steven Rendall (Berkeley, CA and London: University of California Press, 2011).

18 Correspondence between José Joaquín Pérez and Minister of Development, July 12, 1900, folio 377, tomo 823, fondo Ministerio de Obras Públicas, AGN.

19 Correspondence between the chief police officer and Minister of Development, March 22, 1894, folio 356, tomo 823, fondo Ministerio de Obras Públicas, AGN.

20 Correspondence between Guillermo Murillo (gardeners chair) and Director of Public Infrastructure, April 22, 1919, folio 147, tomo 857, fondo del Ministerio de Obras Públicas, AGN.

21 Correspondence between Casiano Salcedo and Minister of Development, September 25, 1890, folio 311, tomo 857, fondo del Ministerio de Obras Públicas, AGN.

22 Correspondence between Casiano Salcedo and Minister of Development, September 12, 1890, folio 311, tomo 857, fondo del Ministerio de Obras Públicas, AGN.

23 Correspondence between Genaro Valderrama and Minister of Economic Affairs (Ministerio de Hacienda), July 22 1895, folio 67, tomo 823, fondo Ministerio de Obras Públicas, AGN.

24 Report made by Genaro Valderrama as Inspector of Parks and Public Gardens, October 8, 1899, folio, 93, tomo 823, fondo Ministerio de Obras Públicas, AGN.

25 The campaign against the *ruana* is comparable to the elites' struggle to eradicate the consumption of *chicha* among the popular classes. All of this social coercion can be considered as steps toward the elimination of the indigenous characteristics still deeply rooted in the society of Bogotá by the end of the nineteenth century. Regarding the *ruana*, for instance, the government passed a bill in 1918 forcing the compulsory use of uniforms rather than that garment for all of the city's carriage drivers, see Rainmundo Rivas, "Labor de la Sociedad," *Boletín de la Sociedad de Embellecimiento de Bogotá*, Mayo 18, 1918.

26 It is important to remember here that the way of dressing was until the beginning of the twentieth century the easiest way to discriminate between social classes.

27 The *fiesta del árbol* was not exclusive of Colombia. Instead, this celebration was undertaken in different Latin American countries, for a Mexican example, see Víctor Talavera, "El Día Del Árbol Durante El Periodo Porfirista En Michoacán 1891–1910/The 'Día Del Arbol!' Throughout the Porfiriato Period in Michoacan. 1891–1910," *Boletín de Antropología* 29, no. 48 (2014): 119–43.

28 Juan Carlos Manó, "El Eucalipto," *El Escolar (Bogotá)*, Octubre 8, 1881, 7–8.

29 The Sociedad de Medicina y Ciencias Naturales (Natural Science and Medicine Society) affirmed about Manó: '*The knowledge on Botany revealed by Mr Manó are inferiors to those presented by any girl in any of the city's school*'. Informe sobre los trabajos del señor Manó, presentado por la comisión a la sociedad de Medicina y Ciencias Naturales', *Revista Médica (Bogotá)*, Diciembre 10, 1882.

30 This section briefly draws on the Foucaultian biopolitics extended to the relationships between people and plants, see Michel Foucault, *Discipline and Punish: The Birth of the Prison* (London: Allen Lane, 1977).
31 Restrepo Zea, *El Hospital San Juan de Dios 1635–1895*, 104–222.
32 Ibid., 132.
33 Lino de Pombo, "Discurso pronunciado por el señor Lino de Pombo, presidente de la Junta de Beneficencia, en el acto de colocar la primera piedra del manicomio de Cundinamarca," *Revista Médica de Bogotá*, Agosto 1913.
34 Adolfo Gómez, *Los Secretos Del Panóptico, Folletines de "Sur América,"* Tomo 1 (Bogotá: M. Rivas y Ca, 1905), 13.
35 Gabriel Castañeda et al., "Informe presentado por la sociedad de medicina y ciencias naturales a la junta general de beneficencia del estado," *Revista Médica*, Julio 20, 1884.
36 Gabriel Castañeda et al., "Informe presentado por la sociedad de medicina y ciencias naturales a la junta general de beneficencia del estado."
37 Grillo, Cortés, and Anzola, *Informe de La Comisión Que Visitó Los Establecimientos de Caridad Dependientes de La Junta General de Beneficencia* (Bogotá: Imprenta Municipal, 1918), 4.
38 Ibid, 11.
39 Restrepo Zea, *El Hospital San Juan de Dios 1635–1895*, 113.
40 Ibid., 126.
41 Grillo, Cortés, and Anzola, *Informe de La Comision Que Visito Los Establecimientos de Caridad Dependientes de La Junta General de Beneficencia*, 8.
42 "Editorial," *Revista Cromos (Bogotá)*, Marzo 1, 1919.

Section III

The trees of Bogotá

8 A treeless and desolate land

The lack of firewood in colonial Bogotá

In 1836, the Scottish explorer James Steuart reached the high-altitude plain of Bogotá. After months of travel through a 'constant succession of beautiful forest', he expressed his surprise at encountering a place where 'the entire want of trees gives the plain throughout an air of great dreariness and desolation.'[1] Indeed, nearly three centuries before Steuart's observations, at the beginning of the colonial period, the anonymous author of the *Relación de Popayán y del Nuevo Reino, 1559–1560* (Report of Popayán and the New Kingdom, 1559–1560), described Santa Fe's [Bogotá's original name] location in a 'very chilly valley, lacking wood, which must be brought from six leagues away from this Spanish settlement'.[2] With some exceptions, the treeless nature of Bogotá was a constant in the descriptions of the city since its foundation to the nineteenth century.

The foundation of Spanish settlements in America was guided by the *Ordinances of discovery, new population and pacification of the Indies* promulgated by Fernando II in 1573.[3] Among many other rules, these laws worked as a manual containing detailed information to be considered by conquerors prior to founding any settlement. In its Chapter 5, the King explains how the new Spanish towns must be located in 'cold places rather than in hot ones' with 'pastures to raise cattle; forest and groves for firewood and construction of houses and public buildings; abundant waters of good quality for drinking and irrigate'. The place also needed to have '*Indios y naturales* (Indigenous and natives) to whom preach the Gospel as it is the primary motivation'.[4] However, as demonstrated in many other cities all over America (e.g. Mexico City or Cuzco), the Spanish superimposed their cities on highly populated native settlements with strong vertical political structures.[5] People to evangelise meant also free human labour to be used to supply local products. Despite the King's guidance it was common that in the creation of new settlements the Europeans privileged the presence of indigenous people over the availability of natural resources. One

DOI: 10.4324/9781003208167-12

of the main consequences of this trade-off between people and resources in Bogotá was systemic problems around the firewood supply for the city.

When translated to the Andean regions, the preference of establishing a town in cold weather prescribed by the distant king meant an endemic shortage of forest products. Despite the existence of some forests characterised by *encenillos* (*Weinmannia tomentosa*), and *ají de páramo* (*Drymis granadensis*) for at least the last 10,000 years, cold conditions have prevented the establishment of dense forests near Bogotá. In addition to this, towards the east, the city is flanked by the steep Cerros Orientales (Eastern Mountains) as well as the *páramos*, a high-altitude, biodiverse moorland barren of trees.[6]

In addition to the ecological conditions, the high native population had also produced a significant impact on the forest of the high plain of Bogotá. Muiscas had largely used the high-altitude forest, which, in contrast with their counterpart in lowland regions, developed and regenerated slower, making it difficult to maintain constant extractive practices.[7] Besides, the indigenous population had systematically replaced native and spontaneous species such as *Drymis granatesis* with domesticated plants. Miguel Daza de Armendariz wrote in 1547 to the Spanish King that within the 32 leagues [1 league = approx. 4.8 km] separating the towns of Velez and Tunja on the northern side of the plain, 'there are not four [leagues] that do not present signals of maize or potatoes culture . . . or beans, cotton or coca'.[8] As I have explained in the Introduction, the Muiscas developed a sophisticated trade network that benefitted from the ecological differences across the Andes. After the domination of Muiscas, the Spanish replaced or integrated this trade network into their encomiendas. Thus, their control was executed using the previously developed indigenous knowledge over their lands (see Chapter 1). This situation is clearly illustrated through the establishment of a vassalage structure through which the Spanish rulers ensured a constant supply of firewood. Known as *Mita de leña*, the scheme forced people from certain *pueblos de indios* to transport specific amounts of daily firewood loads into the city.[9]

One of the most significant constraints faced by the *Mita de leña* scheme was the historical lack of large trees in the mountainous environments near Bogotá, which forced the indigenous people to undertake large trips to obtain the demanded firewood. For instance, records in 1593 reveal how the indigenous people of Teusacá [currently La Calera] travelled 24 km on foot to deliver firewood to Bogotá.[10] However, it seems that the middle-elevation woodlands beyond the Easter Hills and around the current towns of Cáqueza and Fómeque had an outstanding role in the supply of wood to the city. For example, the ledgers of one of Bogotá's most important *Mitayo de leña* [person in charge of the *Mita de leña*] document how during the years 1616–27, he obtained thousands of firewood loads coming from the indigenous town of Ubaque, located 35 km away from the city.[11] The large distance between the woodlands and the city, the indigenous population's decline, and the permanent prohibition on native people's ownership of beasts of burden resulted in a constant renegotiation of the *Mita de leña*. The firewood supply scheme became less stable and rather inefficient over the years, with periods of excessive availability alternating with others of profound scarcity.[12]

Despite the troubles linked to the lack of close forest, the consolidation of the colonial and republican city led to the diversification of many daily life activities relying on a steady firewood supply. An example of this was the wood consumption in the production of *aguardiente*, one of the most important alcoholic beverages of Bogotá. The *aguardiente* was produced through the distillation of the sugarcane juices, which were later flavoured with anise. In contrast to *chicha* [the traditional alcoholic beverage of the native Americans made through the fermentation of maize grains], the production of aguardiente requires heat and, therefore, wood. In a report made by Isidro Lazzo, the outgoing manager of the Real Estanco [Royal Monopoly, a term used to denote the Spanish crown's control over various commodities], to his successor, Miguel de Galves in 1770, the former enumerated all the raw-materials purchased to produce the *aguardiente*. The list included 17,475 earthenware pitchers and one *totuma* of honey, 1,030 arrobas and 8½ pounds of anis, and 17,177 and a half *carga* of firewood.[13] Although the time period during which this firewood was obtained is unspecified, the liquor factory's high demand clearly had large implications on the consumption of the forest resources beyond Bogotá's mountains, and therefore a transformation in the out-of-sight plantscapes of the colonial city.

The city's consumption of firewood increased with the complexification of urban dynamics in the nineteenth century. For instance, by 1868, the Colombian Mint used a considerable amount of firewood in water distillation to obtain something named 'normal solution' used in the elaboration of coins.[14] However, beyond punctual activities such as in the case of the national mint, the city's kitchens were the final destination of a large number of trees turned into logs used for cooking. As late as 1870, the 40,883 inhabitants of Bogotá relied on firewood or coal for cooking their meals, and their demand was supplied by indigenous people who sold sticks and trunks in the so-called *mercados de maderas* (wood markets). Although it is impossible to accurately track the amount of wood consumed by each household, the continuous cutting of public trees and the existence of this market suggest that the circulation of this resource was highly intensive, with potentially large repercussions on Bogotá's nearby plant communities.

Ongoing requirements for firewood and coal in Bogotá led to the systematic exploitation of these resources, which required a continuation

Figure 8.1 Anonymous, *Men carrying a cargo of wood on his back*, ca. 1895, photograph, 11.4 × 16.6 cm.

Source: Dubail, C., and M. Dubai-Acero, eds. *Bourgarel, le Colombien: Voyages d'un diplomate français dans la Colombie du XIXe siècle*. Paris: EdiSens, 2017. Courtesy of the Association Ernest Bourgarel

of the extractive dynamics established with the *Mita de leña* scheme into the late nineteenth century. However, following the colonial period, the people responsible for the wood supply were not the Muiscas, but rather their descendants, impoverished rural-dwelling mestizos [*campesinos*], who continued this traditional pattern of subsistence (see Figure 8.1).[15] As was the case for the people trading NTFPs, those in charge of selling firewood had to face the arduous conditions presented by the wild mountain environments close to Bogotá.[16] In that context, by the end of the nineteenth century, whilst some highly educated Bogotanians were making plans to recreate a European-like nature in the city through the establishment of parks, a larger number of people worked daily in the forests and *páramos* to supply the basic needs of a city that, despite conceiving of itself as modern, was strongly attached to the indigenous cultural roots developed in colonial Bogotá.

By the end of the nineteenth century, city growth put more pressure on forest products. As a consequence, the educated elites introduced large forestry schemes in the city, based on introduced species. Using the case of eucalyptus in Bogotá, Chapter 10 analyses this process in depth. But first, the following Chapter examines how the lack of ornamental trees influenced early attempts to use arboreal elements as decoration and how trees had a significative role in the plethora of symbols associated with the independence from Spain starting in the nineteenth century.

Notes

1 Steuart, *Bogotá in 1836–7: Being a Narrative of an Expedition to the Capital of the New-Granada and a Residence There of Eleven Months*, 107.
2 Anónimo, *Relación de Popayan y Del Nuevo Reino de Granada, 1559–1560* (n.d.), 92–122. In: Patiño, VIctor, *Cespedecia*, 45–6, suplemento no. 4, (January - June, (1983): 23–103.
3 Silvia Arango, *Historia de La Arquitectura En Colombia* (Bogotá: Centro Editorial y Facultad de Artes, Universidad Nacional de Colombia, 1991), 87.
4 Título Quinto, *De las poblaciones, Ordenanza 34.35 y 36 in: El orden que se ha de tener en descubrir y poblar: transcripción de las ordenanzas de descubrimiento, nueva población y pacificación de las Indias dadas por Felipe II, el 13 de julio de 1573, en el bosque de Segovia, según el original que se conserva en el Archivo General de Indias, Sevilla* (Madrid: Ministerio de la Vivienda. Servicio Central de Publicaciones, 1973).
5 Romero, *Latinoamérica Las Ciudades y Las Ideas*, 46–68.
6 Néstor García, Orlando Vargas, and Yisela Figueroa, *Los Cerros Orientales y su flora el Acueducto de Bogotá, sus reservas y su gestión ambiental* (Bogotá: Acueducto de Bogotá, 2006); Pilar Angélica Gómez-Ruiz, Roberto Lindig-Cisneros, and Orlando Vargas-Ríos, "Facilitation among Plants: A Strategy for the Ecological Restoration of the High-Andean Forest (Bogotá, D.C.—Colombia)," *Ecological Engineering* 57 (August 1, 2013): 267–75, https://doi.org/10.1016/j.ecoleng.2013.04.049.
7 Additionally, the Andean Forest grows and regenerates slower than their counterparts in low-land regions. This means that it is, see John Ewel, "Tropical Succession: Manifold Routes to Maturity," *Biotropica* 12, no. 2 (1980): 2–7, https://doi.org/10.2307/2388149.
8 Juan Freide, *Documentos Inéditos Para La Historia de Colombia (1509–1550)*, vol. VIII (Madrid: Arco, 1955), 308; Langebaek, *Mercados, Poblamiento e Integración Étnica Entre Los Muiscas Siglo XVI*, 58.
9 Cristina Vanegas, "Los Pueblos de Indios y El Abasto de Leña de Las Ciudades de Tunja y Santafé, Siglos XVI y XVII," *Fronteras de La Historia* 20, no. 2 (2015): 92–122.
10 Langebaek, *Mercados, Poblamiento e Integración Étnica Entre Los Muiscas Siglo XVI*, 99.
11 Quadernos de pago de leña que hace Pedro de Vergaño, 1616–27.35, D.3, folio 10–24, Sección Colonia, fondo Caciques e Indios, AGN.
12 Vanegas, "Los Pueblos de Indios y El Abasto de Leña de Las Ciudades de Tunja y Santafé, Siglos XVI y XVII."
13 A half *carga* of firewood being a *carga* the equivalent to a load of aprox. 160 cm of circumference, Consumo de mieles, leña y anises en el Real

Estanco de Santafé, 1770, Sección Colonia, fondo misceláneas: SC 39, folios 515–18, AGN.

14 Talones de Vales, Septiembre 1868. Placed in Db 306. Archivo Histórico Casa de la Moneda. Banco de la República, Bogotá, Colombia.

15 Guiomar Dueñas, "Sociedad, familia y género en Santafé a finales de la colonia," *Anuario Colombiano de Historia Social y de la Cultura*, no. 21 (January 1, 1993): 63–113.

16 It is important to mention that until the first decades of the twentieth century, forests and other wild places were thought to be inhabited by demons. The Romantic idea of nature as a representation of divinity reached Colombia late as a result of its 'delayed' modernisation, see Pérez, *La Obra de Dios y El Trabajo Del Hombre: Percepción y Transformación de La Naturaleza En El Virreinato Del Nuevo Reino de Granada.*

9 The Alamedas and the Liberty tree

The alamedas of Bogotá

To underpin the desolation of Bogotá's landscape, the already-mentioned Scottish explorer James Steuart underscored the presence in the city of a line-planted 'fungous sort of willow; shaped like the Lombardy poplar'.[1] These trees made part of one of the two *alamedas* (promenades) of Bogotá. They had been planted as an early attempt to endow the city with an organised urban nature and had an ephemeral life span in the city.

The systematic planting of trees as urban facilities occurred in European cities as early as the sixteenth century; however, this trend did not arrive in Latin America until the eighteenth century.[2] Although many different species of trees, such as peaches, apples or cypresses, arrived with newcomers during the sixteenth and seventeenth centuries, these trees were planted in the rural-like environment in which settlements developed. Until the early eighteenth century, we could assert that there was not such a thing as urban trees. This rural-related context of trees in Bogotá began to change after the War of Succession in Europe (1701–1715). This war gave the Bourbon dynasty control of Spain and its colonies and opened a period of radical transformations called the 'Bourbon reforms'. One of the main consequences of the change of regime in today's Latin America was the expansion of the royal domain through the foundation of thousands of small settlements across the continent.[3] In parallel with the increase in settlements, the Bourbonic administrators undertook concrete urban transformation encompassing improvements in the hygiene, recreation, transportation and economic efficiency of the city.[4] As part of these reforms, and in tune with the enlightened spirit of the eighteenth century, urban embellishment acquired an unprecedented relevance. The ornamentation of urban centres in Latin America crystallised through the creation of *alamedas* and *paseos*, which were tree-lined streets that tried to mimic the promenades planted in European cities (see Table 9.1).

DOI: 10.4324/9781003208167-13

Table 9.1 Main alamedas built in some Latin American cities during the eighteenth century

Name	*City*	*Date*
Alameda	Quito	1746 (refurbished in 1785)
Paseo de Aguas	Lima	1770–1776
La Cañada	Santiago	1762 (Refurbished)
Las Tajamares	Santiago	1770 (Destroyed by flooding)
Paseo del Paula (Pedestrians)	La Habana	1770s
Paseo del Prado (For carriages)	La Habana	1770s
Paseo Nuevo (Paseo Bucareli)	Mexico City	1770
Paseo de las Vigas	Mexico City	1770
Passeio Publico	Rio de Janeiro	1779–1783
Paseo de Agua Nueva	Bogotá	1747
Alameda Vieja	Bogotá	1789
Alameda Nueva	Bogotá	1790

Sources: Lawrence, *City trees*, 96–99.

The echoes of the urban Bourbonic reforms also reached Bogotá. In 1747, the Viceroy José Manuel Solis Folch de Cardona, built an aqueduct to update the city's water supply. The main part of this aqueduct was formed by a canal aiming to take water from the San Francisco River in a place called El Boquerón (depicted in Figure 3.1) and transport it to the fountain located in the Main Square. Once there, the people from the city were able to pick up the water from the mountains using a bamboo stick of chusque (*Chusquea scandens*) and clay pots or containers made of totumos (*Crescentia cujete*), as presented before. The chronicler Pedro Ibañez observed how the canal was 'beautified with the Paseo de Aguanueva (New Water Promenade)' which, according to him, 'competed with the Pincio promenade in Rome'.[5] On May 30, the water from the new aqueduct started to flow after elaborated ceremonial protocols that included a 'dwarf and a monkey as a gift for his excellency [the viceroy]'.[6]

We ignore what types of trees were used in the creation of this early promenade in Bogotá. Nevertheless, we know that the Aguanueva Aqueduct was in use until 1805 when an earthquake destroyed it completely.[7] The promenade remained, however, and rather than being used as a place for walks as in Europe, it turned into a transit space for people bringing different kinds of products such as poultry, charcoal, firewood and fruits to the city. In 1882, the British woman Rosa Carnegie-Williams described how, during her visit to the Aguanueva, the promenade 'was crowded with indigenous women who brought their goods and fruits to the market on Friday'.[8] At the end of the century, the constant confluence of people, usually from marginal backgrounds, promote the emergence of a steady settlement of people living in shanties known as *ranchos*. Thus, during the first decades of the twentieth century, and after being a distinctive sign of enlightened urbanism, the Aguanueva promenade became one of the first slums in Bogotá.[9]

After almost 50 years, another promenade was laid out in Bogotá. In 1789, Don José de Ezpeleta took power as the Nueva Granada viceroy. Ezpeleta had acquired a good reputation as a diligent and enlightened man as Governor in La Habana, where he had led urban improvements in the cemeteries and street lights.[10] Soon after his arrival in Bogotá, Ezpeleta introduce considerable changes in Bogotá such as the creation of a police department and the pavement of previously moody streets with bricks. Additionally, he ordered the construction of the Paseo de la Alameda. Later on, renamed as *Alameda Vieja* (Old Alameda), this promenade ran from West Street until the site called *Puente Aranda* (see Figure 9.1). Some years later, the same Ezpeleta ordered the creation of another promenade that came to be recognised as *Alameda Nueva* (New Alameda). Ornamented also with trees, this promenade ran from San Victorino Square to the countryside beyond San Diego.[11]

The alamedas worked as the city entrances. As all cities built by conquerors in the Andean region, Bogotá did not have any walls around them (the indigenous enemies were easily subdued and, more importantly, in the high-altitude cities, there were no pirates!). Thus, compared with many European cities where the main entrances turned into ceremonial arches as a reminder of the city borders, what welcomed Bogotá's visitors were two lines of trees stretching for several kilometres beyond the town. The alamedas were a sign of 'civilization': the extension of urban life stretching out over the countryside. Both the Alameda Nueva and the Alameda Vieja were planted in parallel to the main two roads that link Bogotá to other major centres. While the Alameda Nueva stretched towards the road that connected Bogotá with the salt-producing village of Zipaquira in the north, the Alameda Vieja stretched over the road that joined Bogotá with the Magdalena River in the west, and from there with the Atlantic world. We need to remember that as late as 1873, reaching Bogotá from England, for example, took a month of transatlantic travel, then ten more days sailing the river Magdalena, and another four days crossing the jungle rising up 2,600 metres to the plateau on which Bogotá is located.[12] In lights of this trip, people who arrived at the city saw the Alameda Vieja as the end of their travel and as a promise of a proper place to sleep, free from the influence of hot weather, mosquitos and snakes.

As a result of their symbolic role as city entrances, the alamedas were considered an integral part of the city. This can be evidenced in the city's cartography produced during the last part of the colonial period, where alamedas are represented as the only evidence of an organised nature with ornamental purposes (see Figure 9.1). However, the inclusion in the graphical representation of the city did not guarantee a full social appropriation of these spaces. Partially because of their geographical position on the outskirts of the city, the alamedas in Bogotá never reached the social importance held by the church atrium as a daily venue.[13] However, in a city with too well-structured life rhythms, and without access to public entertainment, any venue was used in some way or another as a place of social representation. Whilst the alamedas were not actively frequented during the week, they acquired relevance among some sectors of the society on Sundays, when daily life rhythms were modified. About the social use of the alamedas, Karl August Gosselman, a Swedish naval officer who visited Colombia between 1824 and 1826, wrote:

> Although the people from the capital do not very much like excursions—in this they are not compared to the rest of the country—they limit their walks to the royal street. However, on Sundays, the Bogotanians parade

Figure 9.1 Vicente Talledo, *Map of Bogotá*, 1810.

Source: Courtesy of the Museo de la Independencia

> through the delights of this avenue [the alameda]. Those who frequent it in greater numbers are young people, who in their horses travel through it at full trot. This custom of riding is very well seen in Bogotá, since the quality of "mejor señor" (Best sir) is meaningless if he does not own a riding horse.[14]

This description reveals how the alamedas were spaces of social display by the elites and also spaces of gender differentiation. Thus, while some young women of the city's elite were taking care of flowering plants in patios, as explained in Chapter 2, some young men were using the space created with trees as a stage where they showed their equitation skills. This fact reaffirms the role of women eminently linked to the domestic/private spheres, which was the opposite of men's social performance, unfolding in the public sphere. The dichotomy between women/private and men/public was well established, especially among some of the most privileged people in the city.[15] Nonetheless, for many other women, the street and open spaces such as the river were common spaces. Washerwomen (*aguateras*), wood, and weed saleswomen had a completely different experience of open spaces in the city and their nature in surrounding areas. For many women in nineteenth-century Bogotá, the open spaces of the city were their place of work, but also places that hid the dangers of potential sexual assaults.[16] Unfortunately, the historical evidence on the use of early public space by women of different social classes is scarce or simply does not exist.

Despite their differential use by men and women of different social positions in the city, it is important to highlight how the alamedas of Bogotá were just tree-planted roads used by many every day. They were places of transit which did not promote engagements with the plants. However, a closer analysis of the type of trees used in the creation of alamedas in Bogotá reveals perfectly how the models that inspired the creation of urban nature in Bogotá had to be adapted to the environmental reality of the city.

According to the Real Academy of the Spanish Language dictionary, the word alameda means a place or promenade cultivated with poplars. Although the aforementioned Carl August Gosselman described the Alameda Nueva as formed by 'leafy poplars whose trunks are linked in an impenetrable net of branches and roses',[17] it seems that, instead of poplar, these alamedas were planted with species that could withstand the chilly weather of Bogotá. The chronicler Pedro Ibañez observed that, in the alamedas, black cherry (*Prunus serotina*), Colombian oak (*Quercus humboldtii*) and alder (*Alnus acuminata*) were the types chosen to accomplish this urban transformation.[18] Particularly important was the use of willows, of which there are two different species in Colombia: the common willow (*Salix humboldtiana*), which is native to Colombia, and the weeping willow (*Salix babilonica*), an introduced species from China. As the weeping willow was only introduced to Colombia at the end of the nineteenth century,[19] the willow used in promenade construction had to be the native *S. humboldtiana*. This willow shares with all the species of the same genus its ability to grow from cuttings, which makes it easy to propagate. This physiological trait was the reason behind the widespread tradition of use of this tree as hedgerows between properties in rural areas. The use of willows in the alamedas of Bogotá can be considered as a natural extension of their use in the rurality of the city. The same can be said for other species used in the construction of the alamedas such as alder and cherry trees. Besides being common elements of the rural world used now as ornamental, alder and cherry trees tend to grow fast from seeds and they do not require much care during the process, similarly to oaks, the other species present in the alamedas according to Pedro Ibañez.

At the beginning of the twentieth century, alamedas did not exist anymore or were profoundly fragmented. The systematic use of ornamental trees as firewood by many, plus the efforts to make their willows look like poplars, led to the decadence of the alamedas (see Figure 9.2). Thus,

Figure 9.2 Alameda in Mosquera, 1919, photograph.

Source: El liberal Ilustrado, February, 1919. Courtesy of the Biblioteca Nacional de Colombia what had been one of the main signs of urban progress under the Bourbon administration ended up being replaced by new ways of expressing urban progress: the tramways. The need for mobility crystallised in the establishment of modern transport systems such as the tramways, whose construction led to the disappearance of both the Alameda Nueva and the Alameda Vieja in Bogotá (see Figure 9.3).[20] With the tramway, the willows, apple

Figure 9.3 Up (Manuel María Paz, *Outlook of the San Victorino and The Boquerón*, 1840? watercolour on paper, 43 × 33 cm. Down) Manuel María Paz, *Outlook of Bogotá*, 1880? watercolour on paper, 55 × 25 cm.

Source: Courtesy of the Colección de arte del Banco de la República trees and oaks that had been used as the raw material for the alamedas were replaced by poles made out of eucalyptus wood. Seen in that way, the essence of trees remained in what had been the alamedas. Trees turned into poles were not elements of ostentation but essential parts of the incipient electric infrastructure that moved people's bodies now governed by schedules and disciplined work. In short, promenades originated in Europe by the end of the sixteenth century, arrived in Latin America during the eighteenth century with the transition of the crown in Spain and became symbols of ostentation and public display. Nevertheless, by the end of the nineteenth century, when having a horse was not seen as a sign of wealth but rather a symbol of rurality, the alamedas as a place for showing equitation skills lost meaning and they were replaced by well-limited parks and public gardens where disciplined interactions between people and plants emerged.

The liberty tree

The history of plants in Bogotá has endowed trees with various symbolic meanings.[21] The already explained uses of trees as hygiene devices or as symbols of urban progress are just two examples of it. There is, however, a short episode in the history of Bogotá's trees related to the genesis of the country that deserves special attention. Namely, the use of trees as part of the plethora of symbols associated with the independence process of Spain in the early nineteenth century.

When Bogotanians woke up to attend mass on Ash Wednesday, 3rd March 1813 [amidst the Spanish efforts to recover power over the rebel colonies], they discovered a willow (*Salix humboldtiana*) planted in the main square beside the water pile. The tree had a Phrygian cap [symbol of freedom] on the top of it and was also decorated with wide ribbons.[22] Nonetheless, this willow was only provisional. The government planned to replace this first tree with another officially planted one. Thus, some days later, the newborn government invited the 'Generous inhabitants of Cundinamarca and the distinguished citizens of Santafé (de Bogotá)' to attend a ceremony aiming to plant a liberty tree. In the *bando* [as was known official information delivered to the public], they specified how this tree was a 'symbol of freedom' and how the citizens needed to see in the tree the symbol of the recently acquired independence from Spain and the 'possession of our imprescriptible right'. The *bando* also highlighted how the tree needed to be seen as the emergence of a new order and added how 'the symbol that is going to be soon inaugurated [the liberty tree] is not a sign of licence to commit any type of crimes'.[23]

As was planned, the President of the Republic in construction, Antonio Nariño, fixed the 29th of April to celebrate a Civil Holiday, which had as a central event the planting of the liberty tree in the Plaza Mayor (later known as Plaza de Bolívar). As the chronicler José María Caballero remembered, this celebration kept with the protocols of colonial times and gathered most of the people from the city. After a heavy early morning rain ceased, two groups of horsemen showed up in the streets in which house façades were bleached for the event. When the civil procession finished (after the route was shortened due to the time lost because of rain), they finally took out the tree from the *Cabildo*, which was the administrative centre of the local government. According to Caballero, the tree was an *arrayán* (*Myrcia* sp.) of almost five meters in height, planted in a stone triangle built in the main square specifically for it and decorated with many cards written with poems referring to liberty. As with the willow, the arrayán was once again decorated with a Phrygian cap.[24] Afterwards, four wood arches and a decorated lantern were placed over the tree to keep it illuminated. The biggest lantern had depicted on one face the image of Jesus Christ, on the other Mary and in the other two the image of the tree and the justice sword.[25]

To complete the whole symbolic ceremony, the planting of the arrayán included a visible demonstration of the order that needed to be respected. Hence, after being planted and decorated, the liberty tree was used as a place of execution. The day before, a 13-year-old mulatto slave named Antonio Bailly had stabbed his master to death after being punished. When the liberty tree ceremony was coming to an end, Bailly was taken out from jail, and, in front of all the attendants, the improvised police forces made him confess his crime, after which he was executed by four soldiers who shot him twice. After this demonstration of power against a slave, a priest, whose surname was Rosillo, delivered a speech from the jail balcony explaining the meaning of the recently gained freedom. After Rosillo's words, people went back home to warm their bodies with a hot chocolate.[26]

The paraphernalia around the liberty tree did not ensure its survival. Less than two months later, on the 18th of June and two days before the celebration of the third anniversary of independence, the tree was cut.[27] One version says that it was a young man, who after his affront, ran away toward the church. Others explained that the act remained unresolved, despite the reward of 200 *pesos* offered to whomever found the offender.[28] The attack on the republican symbol was accompanied by the appearance of denigrating words towards the president, showing how that the destruction of the arrayán tree was a political protest. However, two days later, the third anniversary of independence was celebrated, claiming (once again) complete independence from the Spanish Monarch Fernando VII the vandalised arrayán was replaced by a Mediterranean olive tree (*Olea europaea*).

As happened across the continent, independence from Spain did not mean whatsoever to break links with Europe. The elites that guided the independence process were in general young and enlightened people inspired by the French Revolution. From France, they got the revolutionary ideas that led to the creation of independent countries and, also from France, they introduced the symbolic universe linked to the replacements of power. The idea of having a liberty tree as a symbol of freedom is one example of this. There in France, Lombardy poplar trees (*Populus nigra* var. *italica*) had been planted everywhere as a celebration of the new social order that followed the suppression of the monarchy in the 1789 revolution.[29] However, when translated to Colombia, this tradition only kept the image of a tree without any specific relation with a particular species. As mentioned earlier, José María Caballero recognised the tree initially planted in the square as a willow, which, according to him, was later officially replaced by an arrayán. However, the priest Torres y Peña, in one allegoric poem addressing the destruction of the first liberty tree, mentioned how this symbol was represented by a completely different species. For Torres y Peña, the first version of the liberty trees was not a willow but a *quina* tree (*Cinchona* sp.), which in turn was replaced not by an arrayán but by a *gaque (Clusia* sp.). Torres y Peña, however, coincides with other sources stating that the definitive liberty tree was an olive tree which lasted until 1816.[30] It seems that, in contrast to France where the liberty tree was closely associated with the poplar, for the recently independent citizens of Colombia, any tree could be used as a liberty tree. In short, as with the gardens planted in the colonial squares or the willows trimmed as poplars in the promenades of Bogotá, the use of an undifferentiated tree to represent liberty reflected how European interaction with plants had to be modified according to the economic possibilities and the social and ecological conditions of this city at 2,600 meters above sea level—a city which in turn was in process of becoming the capital of an independent country.

Notes

1 Steuart, *Bogotá in 1836–7: Being a Narrative of an Expedition to the Capital of the New-Granada and a Residence There of Eleven Months*, 107.
2 Henry W. Lawrence, *City Trees: A Historical Geography From the Renaissance Through the Nineteenth Century* (Charlottesville and London: University of Virginia Press, 2008).
3 Fabio Zambrano and Olivier Bernard, *Ciudad y territorio: El proceso de poblamiento en Colombia, Travaux de l'Institut Français d'Études Andines 64* (Bogotá: Tercer Mundo Editores, 1993), 57–62.
4 Adriana Álzate, *Suciedad y orden: reformas sanitarias borbónicas en la Nueva Granada 1760–1810* (Bogotá: Universidad del Rosario, 2007).
5 Ibáñez, *Las Crónicas de Bogotá y de Sus Inmediaciones*, 191.
6 Ibid.

7 Ibid., 193.
8 Rosa Carnegie-Williams, *A Year in the Andes; or, A Lady's Adventures in Bogotá* (London: London Literary Society, 1882), 103.
9 Mejía, *Los Años Del Cambio*, 389.
10 José Vergara and José Gaitán, *Almanaque de Bogotá, Guía de Forasteros Para 1867* (Bogotá: Imprenta de Gaitán, 1866), 101.
11 Pedro Ibáñez, *Las Crónicas de Bogotá y de Sus Inmediaciones* (Bogotá: Imprenta de la Luz, 1891), accessed June 12, 2016, www.banrepcultural.org/node/70486.
12 Rondy Torres, "Tras las huellas armoniosas de una compañía lírica: La Rossi-D'Achiardi en Bogotá," *Revista del Instituto de Investigación Musicológica "Carlos Vega"* 26, no. 26 (2012): 173.
13 Mejía, *Los Años Del Cambio*, 193.
14 Karl August Gosselman, *Viaje Por Colombia: 1825–1826, (Archivo de La Economía Nacional; 41)* (Bogotá: Banco de la República, 1981), 291.
15 Beatriz Castro, ed., *Historia de La Vida Cotidiana En Colombia, Colección Vitral* (Bogotá: Grupo Editorial Norma, 1996).
16 The perils faced by marginal women in streets and open spaces are usually available in Judicial archives. An example of this is the case of Ana Julia Osorio. A house servant was raped by two men in an unurbanized area named a *rastrojo* in the city of Medellín in 1920. Archivo Histórico Judicial de Medellín (AHJM).
17 Gosselman, *Viaje Por Colombia*, accessed April 2, 2017, www.banrepcultural.org/blaavirtual/historia/viajes/viacolom16.htm. At this point is important to highlight the difficulty that is involved in the reconstruction of some environments and their plants using archive sources. In this case particularly, how to be sure of what Gosselman said about the poplars? Did he have a trained eye that allowed him difference straightforward one specie over other? That is the importance of the dialog between different natures of sources.
18 Ibáñez, *Las Crónicas de Bogotá y de Sus Inmediaciones*, 1891.
19 Diego Molina, *Los Árboles Se Toman La Ciudad. El Proceso de Modernización Y La Transformación Del Paisaje En Medellín, 1890–1950* (Medellín: Editorial Universidad de Antioquia, 2015); Based in: Tomás Carrasquilla, *Medellín* (Medellín: Universidad de Antioquia, 1995); Tomás Carrasquilla, *Frutos De Mi Tierra, Serie Roja* (Bogotá: Alfaguara, 2008).
20 About this topic, see Mejía, *Los Años Del Cambio*, 144.
21 The power inside trees (our perception of them) produce a limitless metaphoric relationship to explain the world as has been nicely expressed by Douglas Davies. Douglas Davies, "The Evocative Symbol of Trees," in *The Iconography of Landscape* (Cambridge: Cambridge University Press, 1988), 32–42.
22 José María Caballero, *Diario de La Independencia* (Bogotá: Banco Popular, [1819] 1974), 130, accessed March 20, 2017, www.banrepcultural.org/blaavirtual/historia/diarioindep/diario6b.htm.
23 Public bando informing about the planting of the Liberty Tree in the central square of Bogotá, 1813, see Biblioteca Digital, Biblioteca Nacional de Colombia, en, accessed April 1, 2022, https://catalogoenlinea.bibliotecanacional.gov.co/client/es_ES/search/asset/88680/0.
24 Trees called as arrayán represent accurately how with the Spanish domination and miscegenation with the Native Americans a new nature with old labels was created. Arrayán is the European name for the trees belonging to the *Myrcia* genus (Eucalyptus family tree), however, that name has been used since conquest times to identify the American *Myrcia* trees.
25 In order to sort the conflict between Federalists and Centralists Pedro Ibáñez, *Las Crónicas de Bogotá y de Sus Inmediaciones*, vol. III (Bogotá: Imprenta de la

Luz, 1891), accessed April 24, 2017, www.banrepcultural.org/blaavirtual/historia/cronic/cap41b.htm.

26 Caballero, *Diario de La Independencia*, 135.

27 Ibid.

28 Vicente Pérez, "El árbol de la independencia en los árboles de la independencia," *Credencial Historia*, ed. 249, September 2010, www.banrepcultural.org/blaavirtual/revistas/credencial/septiembre2010/arbol.htm; Ibáñez, *Las Crónicas de Bogotá y de Sus Inmediaciones*, 1891, accessed April 29, 2017, www.banrepcultural.org/blaavirtual/historia/cronic/cap41b.htm.

29 About the liberty trees, see Lawrence, *City Trees*, 135–37.

30 Ibáñez, *Las Crónicas de Bogotá y de Sus Inmediaciones*. The original source is: E. Lefranc, *Histoire de France,* XI (Paris: Jacques Lecoffre), 573.

10 Eucalyptus
The all-solution tree

The difficulties in finding adequate trees to fit into the emerging cannon of urban ornamentation and national symbolism were a something minor when compared with the increased need for wood, lumber and firewood experienced by the city in the last part of the century. The population increase and the complexation of the urban life experience in Bogotá demanded an increasing supply of wood. However, in spite of the advances in the botanical field that had occurred even before the independence, the information about suitable species to cover the wood demands was scarce. By the end of the nineteenth century, the forestry industry was underdeveloped. As a result, the supply of wood for Bogotá rested on extractive processes from forests on the warmer flanks of the mountains, just as it had in pre-Colombian and Colonial times. Under these circumstances and despite the enormous diversity of the Andean flora, the solution came from abroad in the guise of eucalyptus. This tree would become one of the most important plant species in the capitalist development of the city and, at the same time, will contribute to making Bogotá an island of unique flora by contrast with the city's sounding which slowly became forestry monocultures.

The introduction and uses of eucalyptus

The expansion of eucalyptus trees during the second half of the nineteenth century was a botanical tsunami of global consequences. From being initially perceived by the early British colonisers in Australia as 'monotonous' and 'sombre', and kept as a botanical oddity in some European botanical gardens,[1] in a matter of few decades, eucalyptus' unique biological characteristics [especially an incomparable accelerated growth] turned them into the quintessence of the newborn forestry industry. The already voluminous 'Eucalyptography' has explained how species of *Eucalyptus* botanical genus represented a solution for the growing wood demands of nineteenth-century global industrialisation. Likewise, scholars have shown how the eucalyptus

DOI: 10.4324/9781003208167-14

upsurge was mainly possible by the proselytism of committed key actors, the most important of which was the German but Australian-based botanist Ferdinand Von Mueller.[2] Later recognised as the 'Eucalyptus prophet' Von Mueller established a large network with other eucalyptus enthusiasts, which based on seeds shipments and the spread of technical knowledge, promoted the introduction of eucalyptus in places such as California, South Africa and the Mediterranean basin.[3] One of the key figures of this network was the French botanist Prospel Ramel. Recognised as the 'Eucalyptus apostle', Ramel was responsible for the introduction of eucalyptus in the African French Colonies and indirectly participated in the expansion of eucalyptus towards Latin America. After three years in Australia in which he formed a lasting friendship with von Muller, Ramel Returned to France in 1857. Infatuated by the beauty of eucalyptus but especially of *E. globulus*, he delivered seed to the Paris-based Society of Acclimation of Plants and Animals of which he was an active member.[4] This society would be the source from which eucalyptus seed would be introduced in Ecuador and Brazil in 1865 and 1869, respectively.[5]

The introduction of Eucalyptus in Bogotá occurred during the same period, and as happened everywhere else, it was mainly possible given the commitment of a single person. Although there is evidence of the simultaneous introduction of eucalyptus,[6] the first systematic introduction and acclimation of *E. globulus* was undertaken by Casiano Salcedo, the same self-taught gardener who had transformed the colonial squares into modern gardens. Interviewed in 1918, the elderly Salcedo recalled how he had obtained a handful of eucalyptus seeds from his friend Miguel Salgar, who had received six ounces of *E. globulus* seeds in Caracas while serving as the secretary of then-diplomat Manuel Murillo Toro [who later served two brief presidential tenures in 1864–1866 and 1872–1874] during his posting in Venezuela. As Salcedo recounted, after receiving the seeds, 'I studied in my books about the eucalyptus, a tree belonging to the *mirtáceae* family. I understood its importance; especially because of its accelerated growth, I had information about the large size that some eucalyptus reached in Australia.'[7] Obtaining a massive germination rate, Salcedo planted the seeds and grew 800 small trees.

The eucalyptus trees planted by Salcedo thrived very well in Bogotá's chilly weather. However, in addition to overseeing the trees' acclimation, he was also obliged to mediate to avoid their destruction. When describing the challenges, he had to overcome to succeed in establishing eucalyptus in the city, Salcedo recounted that:

> The first small trees that I planted in "La Pila Chiquita" were pulled out during the night because a gossip said eucalyptus trees were unhealthy. In fact, it is the opposite; they purify the air and they are a healthy

> element against diseases such as malaria. Later, I planted [trees] in "La Estanzuela," but they [the residents] continued with their fight because they believed [the eucalyptus] absorbed all the water from the ground and removed all the nutrients from it. I gave young seedlings to some peasant friends, but shortly they pulled them out because somebody told them that, as it grows, the trees produce an insect that is very venomous and transmits nasty diseases. In the cemetery, there was a eucalyptus with a wrinkled trunk, and someone spread the story that the tree had been infected by a contagious disease and that the tree sweated blood.[8]

Salcedo's efforts ultimately paid off. In 1875, the Frenchman Édouard André, a well- recognised landscape architect and plant hunter, arrived in Bogotá.[9] During one of his trips through the Sabana, he visited the Hacienda Canoas, where, as he later wrote in his travelogue,

> Don Pepe [the owner] showed me a great eucalyptus globulus that, even though it was no more than three years old, it had a fifteen meters height. It could be claimed that the culturing of this famous *mitácea* will leave large fruits in the Bogotá's weather.[10]

André's words proved prophetic. By the first decades of the twentieth century, the eucalyptus had become omnipresent in the Bogotá landscape.

As the city's landscape began to be transformed by the presence of eucalyptus trees, Casiano Salcedo's initial efforts were boosted by the enthusiastic work of Antonio Izquierdo. Having accumulated a large fortune through the newly emerging real estate market, Izquierdo became one of the most influential figures of the Colombian economic elite and a pioneer of the Forestry industry when foreseeing the massive revenues that could be obtained through trees, he began planting several types of eucalyptus in Bogotá in 1894.[11] Thus, the Sabana de Bogotá that Steuart had described as a place of 'great dreariness and desolation' had been significantly transformed by the end of the nineteenth century. However, the efforts of human characters such as Salcedo or Izquierdo would never have been successful if it were not for the intrinsic biology of this tree, which with a high level of acclimation and an accelerated growth cycle turned eucalyptus into a 'worker tree' able to address a large range of needs emerging with the modernisation of the city.

Eucalyptus as a raw material

The central role of the eucalyptus trees in Bogotá was pointed out in the aforementioned interview with Casiano Salcedo. Explaining the historical importance of the eucalyptus introduction to the city, Gabriel Ortiz highlighted that:

> [eucalyptus] which nowadays is part of our landscape, provides us with warmth through its wood; from the hovel to the factory. It bakes the people's daily bread and transmits ideas in the form of telegraph and lighting pole. . . . In the Funza highlands [the eucalyptus] increasingly represents wealth and is an indispensable sign of civilization. Especially the Bogotanians like [the eucalyptus] because it is nowadays a familiar tree, and decoration of farms, avenues and parks name after the father of the country.[12]

Two decades after Salcedo had planted the first set of trees in Bogotá, the eucalyptus had become established as an urban tree and one of the most important raw materials used in the physical construction of the city. The historical lack of trees linked to the extended exploitation of the nearby forest in a region characterised by its slow forest recovery rate had made the warmer sides of the mountains the only sources of wood able to cover the city's demand.[13] As an efficient and highly adaptable 'non-human worker', eucalyptus was viewed as the best solution for the emergent urban demands, and the trees' proliferation substantially reduced Bogotá's dependence on wood from distant locales.

One of the main uses given to the eucalyptus was in the construction of a modern infrastructure. The trees were particularly important in the establishment of Bogotá's tramway network. During the first decades of the twentieth century, the Tram Company had already introduced eucalyptus as a way to reduce operational costs in building and maintaining the city's tramway network. For example, the company spent 2,912.63 Colombian *pesos* on sleepers made from eucalyptus wood in 1912.[14] A year later, after restructuring a company, the new Tram Administration Board declared that they would need as many as 21,550 eucalyptus sleepers due to the poor condition of the tram network they had inherited from the previous administration.[15]

However, eucalyptus was not only used to produce sleepers for the tram system. Since its establishment in 1884, the city's trams had been powered through the living energy of mules, oxen, and horses, thus earning the name 'blood trams'. By 1915, the Tram Company had begun introducing steam-powered carriages, which required a cabling network supported by poles to supply electricity. As a consequence, a complete set of poles was established in all five lines of Bogotá's tramway system. An inventory of the Tram Company's expenses for December 1918 shows that they had spent 178.18 *pesos* in buying and transporting five eucalyptus poles during that month.[16] Given their enormous size, eucalyptus trees were systematically used to construct electricity poles until at least the 1930s (see Figure 10.1).

Figure 10.1 Gumercindo Cuellar, *Jiménez de Quesada Avenue* [see the Eucalyptus-made pole], Bogotá, 1930, negative, 9 × 15 cm.

Source: Courtesy of the Biblioteca Luis Ángel Arango with authorisation of Mario Cuellar

The Tram Company's uses of eucalyptus extended beyond its exploitation as a raw material in the creation and maintenance of the network. In order to ensure a continuous supply of coal for the steam-powered carriages, the company had a mine on a piece of municipal land to the south-east of Bogotá in the Cerro de Guadalupe (Guadalupe Mountain). By 1913, the company had planted 2,000 eucalypti on this land parcel, and according to its report for that year, they were

> working in fencing a space to plant around 30,000–40,000 eucalypti and other tree species to help with the sanitation and the embellishment of the city. Moreover, in the future, those trees will be very important as pit props in the exploitation of the mine as well as for the tram system.[17]

It is not clear if the goal of 40,000 trees was ever achieved; however, by 1915, the land had already been fenced, and another 2,000 trees were planted that year.[18] Thus, the use of eucalyptus as pit props by the tram company reveals how its plantation was established as a kind of 'organic quarry' from which areas around the city were used to cover urban wood demands.

Although eucalyptus supported the whole functioning of the tramways by 1915, the Tram Administrative Board began to look for an alternative wood source. In its annual report for that year, they highlighted that although 'most of the wood used as sleepers came from eucalyptus—due its long life, resistance, availability and low price', they were also interested in using sleepers made of *cúmula* (possibly *Aspidosperma polyneuron*) and *guayacán* [*non-determined species*], emphasising the intention to 'increase the use of other species in future years'.[19] The plan to gradually replace the use of eucalyptus from nearby plantations with high-quality wood from native species spontaneously growing in lowland humid suggested what had been dramatically proved in other regions, namely the *E. globulus* as an unsuitable raw material in infrastructure projects linked to tramways. Already in 1871, the worldwide expert Ferdinand von Mueller had unadvised the use of the blue gum as part of the Australian rail network. In his own words, the 40,000 sleepers demanded by the network on an annual base should not be made out of blue gum mainly because the span did not surpass nine years.[20] The expert's opinion regarding blue gum proved to be right as bitterly discovered something more than a decade later by J.D. Scupham, the assistant chief engineer of the Central Pacific Railroad in California. After planting more than one million trees aiming to be used as sleepers, posts, poles and firewood, soon he discovered that blue gum sleepers would crack if not seasoned properly, and moreover, they rooted away easily.[21] By the beginning of the twentieth century, the bad quality of the blue gum was well known. In a 1908 plants' catalogue produced by the California Nursery Company they clearly stated that the blue gum wood 'soon decay in the ground' making these trees inadequate as raw material for sleepers and poles.[22] Nevertheless, this information was either ignored or unattended in Bogotá where blue gum was at the core of the tramway system development.

Eucalyptus as an ornamental tree

The modernisation of Bogotá involved other elements in addition to the new transport infrastructure. Following the example of the Haussmann Renovation of Paris, facilities such as parks and planted avenues rapidly became synonymous with modernity. Beginning in the 1880s, colonial squares were transformed into small gardens surrounded by iron fences in imitation of the gardens seen by the newborn elite class during their travels through Europe.[23] Despite the enormous diversity of plants thriving in the Tropical Andes, the most biodiverse ecosystem on earth, flowering plants were commonly imported from Paris or New York.[24] Similarly, although native tree species such as *amarrabollos* (*Meriania nobilis*), *siete cueros* (*Tibouchina lepidota*) or *palmas de cera* (*Ceroxylum quindiuense*) were marginally used, by the end of the nineteenth century, the arborisation process extensively relied on introduced species such as pines, araucarias, and especially eucalyptus.

The extensive use of eucalyptus as the default urban tree in Bogotá largely responded to its accelerated growth cycle. The fact that these trees grow faster than any other Andean species places them as an efficient solution to the arborisation demands in a city whose economic constrictions had prevented the implementation of technologies such as those used by the engineers working for Haussmann which had enabled the Parisians to uproot trees and plant them in other places to create well-shadowed promenades from scratch.[25] As Ernst Röthlisberger, a Swiss citizen who the Colombian government had hired to run the Philosophy and History Department at the Universidad Nacional (National University) observed in 1881,

> It is amazing to see how fast trees grow on the city parks. It is important to point out that in Bogotá and its surroundings eucalyptus are specially planted, because of its leafiness and because in a few years they reach high height.[26] (See Figure 10.2).

Fifteen years after its introduction to Bogotá by Casiano Salcedo, the eucalyptus had become one of the most abundant trees in the city not only due to its extensive use as raw material but also because of its value as an ornamental tree. In 1881, the Englishwoman Rosa Carnegie-Williams highlighted the eucalyptus as a tree profusely planted in public spaces:

> a boulevard made of eucalyptus runs along the square, parallel the 5th street, in the west. The eucalyptus trees, that are excellent, were the first to be planted in Bogotá, just fifteen years ago. They are better known as eucalyptus globulus. They adorn the entire city and nowadays have been planted thousands around, given a nice aspect to the plain scare of trees.[27]

Photographic evidence confirms the ubiquitous presence of eucalyptus in the city. Images from this period reveal eucalyptus growing not only in parks but also in a whole range of urban spaces that included the open spaces nearby churches and small chapels, alongside unpaved streets and abandoned plots or growing isolated in the Monserrate's foothills in the east of Bogotá (see Figure 10.3). Very suggestive is the evidence of eucalyptus in solares in the poor and still rural-like neighbourhoods of Bogotá. It is likely that, as still occurs in peri-urban areas around the city, eucalyptus had become an important species for the poorest residents, who found in this tree a partial source of firewood to supplement the reduced numbers of native trees growing around their houses such as the *laurel de cera* (*Morella parvifolia*).[28]

Figure 10.2 *E. globulus*-made promenade in the Parque Centenario in Bogotá, 1883, photograph.

Source: Dubail, C., and M. Dubai-Acero, eds. *Bourgarel, le Colombien: Voyages d'un diplomate français dans la Colombie du XIXe siècle*. Paris: EdiSens, 2017. Courtesy of the Association Ernest Bourgarel

Decline of a trend

After being cultivated for around 40 years, *E. globulus*' nearly three-meter annual growth rate became problematic.[29] The trees' height was overwhelming for a city that had recently begun to transform its physical colonial legacy at the beginning of the twentieth century. Therefore, applications seeking to obtain authorisation to cut down trees began to arrive

Figure 10.3 Hamilton Wright, *Pilgrimage to Monserrate* [Isolated Eucalyptus tree recognised because of the detachment of its bark], ca. 1910, photograph, 20 × 25 cm.

Source: Courtesy of the Biblioteca Pública Piloto

at the Concejo Municipal (City Council). As the highest level office of the city administration, in addition to dealing with the most important urban concerns, the council also had to authorise the cutting of trees (see Figure 10.4).

One of the most common reasons provided for requests to cut down the trees was related to their inconvenient size. The adverse effects produced by eucalyptus were already evident by 1920, as evidenced in a conflict between the Superintendent of the Bogotá Electric Company and Delio Cifuentes, the Director of Public Works. The Superintendent asked Cifuentes to 'give the necessary orders to trim the trees in the Paseo Bolívar . . . since they are rubbing the electricity conduction cables causing troubles in the area'; however, Cifuentes refused on the grounds that

> This department has explicit orders made by the Sr. Minister forbidding the trimming of trees placed in the national avenues. And, to avoid the

Figure 10.4 Daniel Rodríguez, *Aserradores* [Sawyers chopping a eucalyptus tree with another one in the background at the Parque de la Independencia], ca. 1920, photograph (negative) 6 × 6 cm.

Source: Courtesy the Museo de Bogotá

rubbing with the electricity wires, it would be convenient that the electricity company would lay taller poles.'[30]

This reply was not welcomed by the electricity company, which retaliated by cutting off the area's electricity service for several days. This action seems to have been effective. Following complaints made by the sector's residents, the secretary relayed the minister's order to 'trim the trees in the Paseo Bolívar and so avoid their interference with the electricity service'.[31] This example reflects a recurrent conflict between trees, usually eucalyptus, and the city's electricity poles, which were in fact commonly made of eucalyptus wood (see Figure 10.1).

The conflict over electric lines was far from the only clash triggered by the eucalyptus in Bogotá. It seems that the eucalyptus tree planted in the Paseo Bolívar [probably in the last decades of the nineteenth century] was the subject of constant complaints by its neighbours. Another example of the eucalyptus *versus* city confrontation occurred in the same year as the electric company's request. On the 5th of March, José Telésforo Puentes wrote to the Director of the National Public Works Department in the following terms:

> I am the owner of the house number 4 in the 25 street of this city. Since in front of my house was planted a eucalyptus tree that has been alarmingly developed with an evident risk to my house, I request you that according to that reasons mandate to cut down this tree that could destroy my house.

In this case, Delio Cifuentes did not require additional persuasion, he gave the order to 'cut the tree and in its place to plant a smaller one, or leave the space empty since in front of the house there is an electric pole'.[32]

The problems posed by the massive height of the city's trees began to be resolved, at least in part, with the creation of the Sociedad de Embellecimiento de Bogotá (SEB) in 1917. Exclusively comprised of well-educated and wealthy men, this private organisation's establishment was a clear manifestation of the official incapacity to deal with some of Bogotá's most important urban issues. During its initial session on April 17, 1917, the Society's objectives were defined and its commissions were organised according to the city's needs. The Society took charge of Bogotá's parks and public gardens through a commission headed by José María Saíz, a businessman who owned the profitable Faenza pottery factory. Another businessman, the aforementioned Antonio Izquierdo, was involved in issues involving urban tree management. One of the commission's first acts was to sign a contract with Jorge Velez, the Minister of Public Works, whereby the national government transferred the administration of the city's parks, avenues, and gardens to the SEB.[33] Thus engaged with the conservation and improvement of Bogotá's five parks, four avenues, and 11 gardens, each society member became an inspector of at least one of these urban green spaces. The SEB hired guard-gardeners to care for the plants while also supervising park visitors' behaviour. In addition, a set of minimal infrastructures was undertaken to improve the maintenance of public spaces.

The establishment of the SEB marked a change from previous attitudes towards the planting process in Bogotá, which thus far had been rather

Figure 10.5 Municipal Nursery established by Bogotá's Embellishment Society, 1918, photograph.

Source: El Gráfico, March 1924. Courtesy of the Biblioteca Nacional de Colombia

haphazard, and it also meant the end of the systematic default use of eucalyptus trees as the 'urban tree' adorning the built environment. The diversification of trees used in the planting process was made possible when SEB opened a plant nursery in 1918 on a piece of land donated by the Tram Company (see Figure 10.5). According to the SEB's bi-annual report published in December of 1919, the nursery already had 1,252 trees of 11 species ready to be planted in the required places. As seen in Table 10.1, none of the trees grown in the municipal nursery belonged to the *Eucalyptus* genus.

During the same period, the SEB reported planting 340 trees in several locations around the city, including 249 *cauchos*, 65 *acacias*, 12 unspecified palms,[34] eight *alcaparros*, six pines, nine Colombian walnut trees, nine little peach trees, and only two eucalypti in the Parque Santander. The figures suggest a *volte-face* in the role of eucalyptus under the Society's supervision. These new administrators of public spaces appear to have chosen the fig as their favourite species to be planted in the city's avenues and streets, although it is not clear how or why this tree became the urban tree *par excellence*.

Table 10.1 Trees in the municipal plant nursery established by the SEB in 1919s.

Vernacular name	*Scientific name*	*Number*
Caucho	*Ficus soatensis*	908
Colombian walnut	*Junglans neotropical*	142
Andean cedar	*Cedrela montana*	42
Cherry	*Prunus serotina*	57
Alcaparros	*Senna viarum*	12
Peruvian pepper	*Schinus molle*	5
Pines	*Pine* sp.	30
Acacias	*Acacia melanoxylum* or *Acacia decurrens*	5
Little Andean peach	*Abatia parviflora* (cf.)	40
Garagallos	A *non-determined species*	4
Avocados	*Persea americana*	4
Jagüitos	*Pachira mutisiana* (cf.)	2
Total		1,252

Having assumed the role of Bogotá's urban tree specialists, the society's members, particularly those involved in the Parks and Public Gardens Commission, administered the planting of trees in the city. As a consequence, they received voluminous correspondence from officials and private developers constantly seeking advice about suitable tree species to be planted in their newly created neighbourhoods. For example, Cundinamarca State requested the Commission's recommendation for trees to be planted in the Avenida Colón, to which Mr Saíz and Mr Pérez replied that

> The commission thinks that the trees in good shape must be left as they are. With respect to the new one, it must be planted those that have been proved to thrive in this city, such as the acacias in the Avenida Boyacá, Colombian walnuts or Peruvian peppers.[35]

The Parks and Garden Commission's opinion on eucalyptus is explicitly revealed in correspondence during August 1918, when the urban developers of the neighbourhood known as El Ricaurte requested their advice about the most suitable trees to be planted in this newly established, working-class sector in the south side of the city. The Society's reply presented the eucalyptus as strongly unrecommended due to the risk that the tree's roots and height posed for streets and constructions. Rather, they advised that the developers plant acacias and figs and provided additional suggestions concerning the distance between the trees as well as strongly advocating that they plant lines of the same species in order to produce an aesthetic outcome.[36]

The above opinions expressed by the SEB reveal how with the creation of this institution, the planting process became regularised and organised according to rational decisions, which in the pursuit of order, beauty and economy deposed eucalyptus as the urban tree *par excellence* after a reign of nearly 60 years. However, that was not the end of the relationship between Bogotá and the eucalyptus. Although eucalyptus trees were discredited as the best option for urban ornamentation by the 1920s, they became essential non-human workers in the transformation of Cerros Orientales (The Eastern Hills), where they were used to control the water currents of the main city's rivers.

A renewed role: eucalyptus as a 'water regulator'

Bogotá was founded between the San Francisco and San Agustin rivers, and urban life was dictated by their behaviour. During the rainy seasons from March to June and September to November, the same waterflows that dragged the filth away placed the city constantly at the mercy of unexpected floods. Meanwhile, during the dry seasons from June to August and December to February, the city's well-known unsanitary constantly brought cyclic epidemic outbreaks. In 1889, a reported 1,338 people in Bogotá died due to infectious diseases. Mayor Higinio Cualla attributed the main cause of these outbreaks to the low level of the city's rivers and 'recognizing the impossibility of undertaking infrastructure improvements due to economic limitations', he ordered compulsory tree planting in the river basins. Although several attempts to control the river's currents through reforestation had previously been undertaken, the 1889 bill was distinct in that the Mayor emphasised some of the biological characteristics desirable in the trees to be used for river basin planting, among which the most important was a rapid growth rate. According to Cualla, rapid growth would prevent the trees' vandalisation and destruction, which had been a chronic problem plaguing earlier arborisation attempts.[37] The determination to plant the mountains with a fast-growing tree shown by Cualla was an invitation to use eucalyptus, which, by that time, had proven to be a 'fast worker tree' in the recently opened parks.

By the end of the nineteenth century, the role of vegetation on water regulation was an accepted fact. In the British Colonies of the Indian Ocean, the emergent engineering guild have used this proven relation between plants, soil and water to create 'artificial forest' to control desiccation in regions affected by the loss of natural vegetation.[38] Following the same engineering paradigm, in an article published in the *Anales de Ingenieria* (Engineering Annals) in 1899, the Colombian engineer Fortunato Pereira defended the use of the eucalyptus as a way to increase riverine water levels because the tree offered 'advantages not seen in other vegetal species'.[39] However,

Pereira's opinion ignored that although eucalyptus plantations had proved to be efficient drainers in marshy areas usually affected by Malaria,[40] their power as water pumps had promoted regulations forbidding eucalyptus plantations close to water sources in dry areas in India and South Africa.[41] Thus, eucalyptus' large water demands made these trees convenient 'workers' for areas constantly affected by flooding, however, this was not the case for arid regions or with seasonal weather patterns where their presence treated the water availability.

In Bogotá, whose bi-modal rain regime was responsible for several months of drought every year, eucalyptus raised concerns among the engineering guild. In 1912, engineer Diodoro Sánchez recommended 30 species of trees suitable to be planted in the Cerros Orientales in order to create the 'big sponge' that the city needed 'to retain waters coming from heavy rains and then release them purified and regularly'.[42] All of the species suggested by Sánchez were good quality, non-native wood trees, and his list encompasses eight gymnosperms (including three species of pines), one acacia, two kinds of catalpa (*Catalpa* spp.), and—overlooking the autochthonous well-known Colombian walnut *Junglands neotropica*—one introduced walnut tree. Although Sánchez highlighted the possibilities offered by other eucalyptus species such as *E. leucoxylon* as a source of good-quality wood, he willingly excluded from his list the broadly used *E. globulus*. Similarly, the engineer Miguel Triana strongly discouraged the use of *E. globulus* for the arborisation of the river basins. Like Sánchez, Triana perceived that the arborisation of the nearby areas would turn them into a kind of biological quarry, and he promoted the multiple uses of trees in the river basins in his 1914 book *La arborization y las aguas* (Arborisation and Waters). Triana envisioned that the plantation scheme in the city's river basins should be able to not only regulate the city's waters but also contribute to the city's construction projects by covering lumber demands and fuel for the production of tiles and bricks.[43]

The reforestation of the river basins also had an undeniable economic dimension, which was clearly expressed in the position of Antonio Izquierdo. Like the engineers, Izquierdo advanced the multipurpose orientation of the city's forestry scheme; however, he maintained a strong defence of planting eucalyptus. One of the wealthiest persons in Colombia in the late nineteenth century, Izquierdo was also a member of the Sociedad de Embellecimiento de Bogotá, and he proclaimed the usefulness of eucalyptus as a means of earning wealth and providing a social contribution as well as a solution to the city's water regulation problems. In 1918, representing the SEB in the *First Conference of Public Improvements*, Izquierdo delivered a speech in which he claimed that after introducing and successfully cultivating more than 50 varieties of eucalyptus and pines around Bogotá, he had 'sold grounds planted with eucalyptus and pines in Chapinero

(Bogotá area) for more than 20,000 dollars per hectare', as well as 'sold lands to the Municipality with a total area of 1,550 *fanegadas* [992 ha] in the San Francisco, San Cristobal and Arzobispo river basins' and added that 'I had planted trees in large parts of these lands in order to supply the city of Bogotá with pure water coming from the surrounding mountains'.[44] In the same tone, in his self-published booklet titled *Studios Sobre Los Bosques* (Studies on Forests), Izquierdo praised the value of the eucalyptus plantations near the city due to 'the constant provision of water for domestic uses, the diminution of flooding, the urban forest parks, and as places to laid tuberculosis asylums' as well as 'a secure business, with outcomes comparable with any other industry'.[45]

Ignoring the Miguel Triana's and Diodoro Sánchez's arguments concerning the disadvantages of *E. globulus*, on April 8, 1918, the reforestation of Bogotá's river basins began in places where 'the overexploitation or the soil aridity, make them unproductive'.[46] According to Alberto Portocarreño, Bogotá's aqueduct manager, 'the city initially only planted eucalyptus trees due to the lack of access to other types of seeds', although, as mentioned earlier, at that time, the SEB already had an active plant nursery with more than a thousand plants. Two years later, Portocarreño reported that from 1918 to 1920, the city had planted 41,247 eucalypti, 69,189 pines, 9,669 *arbolocos* (Smallanthus pyramidalis), 1,908 cedars, and 12 Colombian walnuts, along with 73,000 plants of the Andean bamboo known as *chusque* (*Chusquea* sp.).[47]

The failure of eucalyptus as a water machine

The establishment of forestry plantations as an organic substitute for the expensive, large-scale engineering infrastructural projects needed to address the water needs of Bogotá's rapidly growing did not mean the end of critics against the use of eucalyptus. In a 1922 letter sent to Fernando Carrizosa, the manager of the Bogotá Aqueduct Company, the renowned naturalist and Lasallian Brother Maria Apolinar claimed that 'The *Eucalyptus globulus* is a contraindicated species when the purpose is to increase the water in a river basin. With this type of tree, the evaporation could reach extraordinary levels'. He recommended 'the use of the species from our cold weathers such as the Andean cedar (*Cedrela montana*), *encenillos* (*Weinmannia* spp.), *raque* (*Vallea stipularis*), Colombian walnut, as well as some exotic species such as araucarias and pines to be acclimated'. Apolinar further explained that trees were unsuitable in areas considerably lacking in deep soils, and—uniquely among his contemporaries—highlighted that leaving spontaneous vegetation to cover the area would be the best solution.[48]

As a response to the arguments against the use of eucalyptus presented by Apolinar, Fernando Carrizosa decided to seek high-level advice concerning the city's arborisation and sent a letter to the United States Forest Service in 1922. Carrizosa addressed some of the issues that had not been solved by the local actors involved in the reforestation process and queried if 'despite its evaporation and absorption power, is this tree recommendable to increase the water flow in this region' as well as whether or not it was necessary to remove the eucalyptus trees that had already been planted in the humid, low-elevation mountain areas.[49] Although Carrizosa's questions accurately reflected the Bogotanians' main concerns regarding the role of trees in regulating water, the reply that he received on December 16, 1922, did not fulfil his expectation of a resolution. Joseph Kittiedge, the United States Chief of Forest Investigation, stated that

> It has been impossible to establish the relationship between the forest cover and water sources, in this or another country. Under the current scientific state, we are unable to give a clear answer to your questions. The studies undertaken so far are contradictive in terms of results.

The North American expert added that

> In our opinion, any local species of tree could be as good as any introduced one to reach your aim and advised that it was uneconomical to replant the eucalyptus trees from the bottom top the top of the mountains.

Interestingly, the lack of clarity in the Kittiedge response ignored the already established knowledge among United States Scientific institutions that had recognised eucalyptus species as 'marsh reclaimer' able to remove large quantities of water from the ground. Such is the case of an 1897 book published by the US Department of Agriculture explaining the use of *E. globulus* in the fight against malaria. Apart from the exhalation of healthy volatile oils, they underscored how its 'rapid growth and the great absorbent power of its roots' endowed *E. globulus* with particular sanity value.[50]

Although any responses to the above brief answers and excuses were not curated, Kittiedge sent a volume titled *Forest and Water on the Shed of Scientific Research*, which Carrizosa used as a textbook. Upon the completion of his study, he concluded that eucalyptus trees were not suitable for the river basins around Bogotá; rather,

> The spontaneous vegetation in the city basins gathers the required characteristic to keep and regulate the water produced by rain. This vegetation produces a compact and absorbent vegetable layer composed by a dense matrix of roots and moss . . . conserving the humidity for long.

Thus, having decided that the native plant species were more appropriate than the introduced ones, the massive reforestation scheme came to its end as the city left the area free for spontaneous regeneration to take place.[51]

As occurred with the low quality of the eucalyptus sleepers, the potential negative consequences as water regulators were unforeseen or ignored and only considered when eucalyptus trees had been extensively used. Thus, 20 years after Bogotá's Mayor proclaimed the need for fast-growing trees in the city's river basins to increase the levels of available water, the 437,600 trees planted between 1918 and 1922 had proved ineffective, as there was no significant change in water currents.[52] Furthermore, they seemed to have long-lasting noxious effects on the soil humidity. In 1924, six years after the beginning of the planting process, Arnold Schultze published an article in the *Revista de Industrias* (Industrial Journal) in which he decried the deleterious consequences brought to Bogotá by the massive plantation of eucalyptus. As an example, Schultze presented an image of the Chapinero sector of the Cerros Orientales, where 'a eucalyptus forest planted seven years ago had dried the soil into a distance of two and a half meters creating deep cracks of a meter and a half deep' (see Figure 10.6). Schultze concluded that eucalyptus trees were eminently unsuitable for places where it is important to maintain soil humidity: 'As occurred in the river known as *Chicó*, where the quantity of water decreased at the same time the eucalyptus plantation have thrived in that sector of the Cerros Orientales'.[53]

The blue gum's legacy

After five decades of being used as a key non-human worker in the provision of raw materials, the city's ornamentation and as a water control device, the eucalyptus euphoria in Bogotá came to its end. Other species and technical-based solutions brought to an end its place as the urban tree *par excellence*. However, the decline of the eucalyptus as a critical working tree in urban functioning did not lead to its disappearance. On the contrary, eucalyptus biological characteristics ensured it a place in other aspects of Bogotá's daily life. Currently, eucalyptus has been culturally appropriated by local communities. People in Bogotá use eucalyptus for different domestic purposes among which the most important is the medicinal.[54] In any traditional market of Bogotá, branches of eucalyptus are sold as an efficient domestic treatment for respiratory minor diseases. Additionally, since its ability to produce high quantities of wood remained unaltered after losing its role in the city, the blue-gum has been adopted by rural and peri-urban communities to supply their needs of wood and fuel.[55] Thus, once thought of as a 'hard worker tree' in service of nineteenth-century modernisation, eucalyptus has become a domesticated plant that, nevertheless, continues to play a significant role in Bogotá where it is a key ecological ally in the regeneration

Figure 10.6 Cracks on the soil apparently produced by the eucalyptus plantation in the Chicó suburb, 1924, photograph.

Source: Revista de Industria, July 1924. Courtesy of the Biblioteca Nacional de Colombia, Bogotá

of the Cerros Orientales' natural vegetation.[56] However, in spite of its apparently key role in the native vegetation regeneration, the still vigorous eucalyptus plantation in the Cerros Orientales hosts less plant species than the urban gardens thus contributing to make by contrast, the urban flora in the built environment of Bogotá richer in species than its surroundings.

Notes

1 Robin Doughty, *The Eucalyptus: A Natural and Commercial History of the Gum Tree* (Baltimore, London: Johns Hopkins University Press, 2000), 27–28; John Wrigley and Murray Fagg, *Eucalypts: A Celebration* (Crows Nest, NSW: Allen & Unwin, 2010), 91–92.
2 Some examples of the Eucalyptography include Brett M. Bennett, "The El Dorado of Forestry: The Eucalyptus in India, South Africa, and Thailand, 1850–2000," *International Review of Social History* 55, no. S18 (2010): 99–59, https://doi.org/10.1017/S0020859010000489; Doughty, *The Eucalyptus: A Natural and*

Commercial History of the Gum Tree; Robert LeRoy Santos, "The Eucalyptus of California," *Southern California Quarterly* 80, no. 2 (1998): 105–44, https://doi.org/10.2307/41171891; Jackson R. Perry, "'Conquered by the Sparrows': Avian Invasions in French North Africa, circa 1871–1920," *Environmental History* 25, no. 2 (April 1, 2020): 310–34, https://doi.org/10.1093/envhis/emz096; Veli Pohjonen and Timo Pukkala, "Eucalyptus Globulus in Ethiopian Forestry," *Forest Ecology and Management* 36, no. 1 (August 1, 1990): 19–31, https://doi.org/10.1016/0378-1127(90)90061-F; Wrigley and Fagg, *Eucalypts*.

3 P. MacOwan, "The late Baron Sir Ferdinand von Mueller, C.M.G., F.R.S., Etc., Government Botanist of Victoria," *Agricultural Journal of the Cape of Good Hope* 9 (1896): 627–28.

4 Doughty, *The Eucalyptus: A Natural and Commercial History of the Gum Tree*, 45.

5 Misael Acosta, *El Eucalipto En El Ecuador* (Quito: Editorial Ecuador, 1947), 7; Edmundo Navarro de Andrade, *O Eucalipto* (São Paolo: Chacaras e Quintais, 1939), 8.

6 The introduction of eucalyptus in Colombia seems to have occurred through different ways. Ernesto Guhl described the Uribe Vanegas family as eucalyptus importers in the 1860s. Additionally, botanist Enrique Pérez Arbeláez, describes how his grandfather imported and planted eucalypti between 1872 and 1882, see Ernesto Guhl, *Colombia: bosquejo de su geografía tropical*, vol. I, 2da ed. (Bogotá: Universidad de los Andes, 2016), 62; Enrique Pérez Arbeláez, *Plantas Útiles de Colombia*, 4da ed. (Santafé de Bogotá: Litografía Arco, 1978), 504–6.

7 Ortiz Williamson, "Casiano Salcedo."

8 Ibid.

9 Stéphanie de Courtois and Florence André, "Édouard André (1840–1911): Cultural and Botanical Exchange between Europe and South America," *Studies in the History of Gardens & Designed Landscapes* 39, no. 3 (July 3, 2019): 178–98.

10 Pablo Navas Sanz de Santamaría and Benjamín Villegas Jiménez, eds., *Colombia en Le Tour du Monde*, vol. I (Bogotá: Villegas Editores, 2013), 675.

11 Izquierdo, *Estudio Sobre Bosques*, 20.

12 Ortiz Williamson, "Casiano Salcedo."

13 Contrasting to the accelerated regeneration of the low-land tropical forest, the cold temperatures present in the Andean forest slow its ecological regeneration, see John Ewel, "Tropical Succession: Manifold Routes to Maturity," *Biotropica* 12, no. 2 (1980): 2–7.

14 Junta Administradora del Tranvía Municipal [hereinafter, JATM], *Informe General Noviembre 1912* (Bogotá: imprenta de carteles), 16.

15 JATM, *Informe General Noviembre 1913* (Bogotá: Talleres tipográficos Régulo Dominguez), 13.

16 B. Plata, "Gastos de conservación, explotación, y administración hechos durante el mes de diciembre de 1918," *Registro Municipal (Bogotá)*, January 31, 1919, 3392.

17 JATM, *Informe General 1913*, 13; The introduction and use of eucalyptus for mining purposes in Bogotá resembles the introduction of *Eucalyptus grandis* in South Africa as key raw material of the diamond mining, see Brett M. Bennett and F. J. Kruger, *Forestry and Water Conservation in South Africa: History, Science and Policy* (Acton, ACT: Acton, ACTANU Press, 2015), 31–35.

18 JATM, *Informe General Noviembre 1915* (Bogotá: Casa editorial de Arboleda y Valencia), 13.

19 JATM, *Informe General Noviembre 1915*, 14.

20 Ellwood Cooper, *Forest Culture and Eucalyptus Trees* (San Francisco: Cubery & Company, 1876), 24–25.
21 Frank William Purdy, *Eucalyptus Cultivation in California, 1853–1900* (Sacramento California State University, 1964); LeRoy Santos, "The Eucalyptus of California."
22 California Nursery Co. et al., *Price List 1908–1909: Fruit and Ornamental Trees Evergreens, Palms, Shrubs, Roses Etc.*, 24th ed. (Niles, CA: California Nursery Company, 1908), 18, www.biodiversitylibrary.org/item/152759.
23 Cendales, "Un parque extenso y amplio para dotar con él a nuestra querida capital," 25–38; María Guerrero, "Pintando de verde a Bogotá: visiones de la naturaleza a través de los parques del Centenario y de la Independencia, 1880–1920," *Revista de Historia Ambiental Latinoamericana y Caribeña* 1, no. 2 (2012): 112–39.
24 Casiano Salcedo and Minister of Economic Affairs, July 10, 1897, folio 189, tomo 823, fondo del Ministerio de Obras Públicas Archivo General de la Nación [hereafter, AGN].
25 Sigfried Giedion, *Space, Time and Architecture the Growth of a New Tradition, The Charles Eliot Norton Lectures 1938–1939*, 4th ed. (Cambridge, MA: Harvard University Press, 1963).
26 Ernst Röthlisberger, *El Dorado: Estampas de Viaje y Cultura de La Colombia Suramericana, Edición facsimilar de la primera edición Colombiana, Apuntes Maestros*, vol. I (1897; reprint, Bogotá: Universidad Nacional de Colombia. Rectoría, 2016), 66.
27 Rosa Carnegie-Williams, *Un Año En Los Andes, O, Aventuras de Una Lady En Bogotá, Colección Viajantes Y Viajeros* (1882; reprint, Bogotá: Academia de Historia de Bogotá, 1990), 73.
28 Fabio Vargas, Fabiola Mejía, and Carlos Guerrero, "Leña como combustible doméstico en zonas rurales de Usme, Bogotá," *Informador técnico* 75 (2011): 30–39.
29 Santelices Rómulo, "Desarrollo de Una Plantación de Eucalyptus Globulus Establecida En Primavera Con Diferentes Tratamientos de Riego," *Bosque (Valdivia)*, no. 3 (2005): 105.
30 Superintendent of the Bogotá Electric Company to Delio Cifuentes, September 20, 1920, folio 109, tomo 816 (Paseo Bolívar), fondo Ministerio de Obras Públicas, AGN.
31 Minister of Infrastructure to Delio Cifuentes, October 5, 1920, folio 111, tomo 816 (Paseo Bolívar), fondo Ministerio de Obras Públicas, AGN.
32 Telésforo Puentes to Delio Cifuentes, March 4–8, 1920, folio 81, tomo 816 (Paseo Bolívar), fondo Ministerio de Obras Públicas, AGN.
33 "Decreto 61 de 1917," *Boletín de la Sociedad de Embellecimiento de Bogotá*, May 6, 1918, 2.
34 This palm could correspond to the wax palm, *Ceroxylum quindiuense,* Colombia's national tree.
35 Minute, March 13, 1918. Libro 2, 090-65, (Book without sheet numeration), Actas de la Sociedad de Embellecimiento, 1918–19, Museo del Chicó, Bogotá [hereinafter, MdC].
36 Minute, August 28, 1918., Libro 3, 090-74, Folios 3–4, Actas de la Sociedad de Embellecimiento, 1918–19, MdC.
37 "Salubridad pública," *Registro Municipal (Bogotá)*, July 5, 1889, 1859–1860.
38 Gregory Barton, *Empire Forestry and the Origins of Environmentalism* (Cambridge: Cambridge University Press, 2002), 99–104; Bennett and Kruger, *Forestry and Water Conservation in South Africa*; Benjamin Weil, "Conservation, Exploitation, and Cultural Change in the Indian Forest Service, 1875–1927," *Environmental History* 11, no. 2 (2006): 319–43.

39 Fortunato Pereira, "El eucalyptus," *Anales de Ingeniería* 9, no. 132 (1899): 221.
40 Charles Thomas Kingzett, *Nature's Hygiene* (London: Baillière, Tindall, and Cox, 1880); Kenneth Thomson, "The Australian Fever Tree in California: Eucalyptus and Malaria Prophylaxis," *Annals of the Association of American Geographers* 40 (June 1970): 237.
41 Bennett, "The El Dorado of Forestry"; Bennett and Kruger, *Forestry and Water Conservation in South Africa.*
42 Diodoro Sánchez, "Árboles que debemos aclimatar," *Anales de Ingenieria* 20, no. 234 (1912): 224–30.
43 Miguel Triana, *La Arborización y Las Aguas* (Bogotá: Casa Editorial de El Liberal, 1914), 1–31.
44 Izquierdo, *Estudio Sobre Bosques*, 10–11.
45 Izquierdo, *Estudio Sobre Bosques,* 20.
46 Alberto Portocarreño, "El gran problema de Bogotá, el acueducto y las hoyas hidrográficas," *El Tiempo (Bogotá)*, Febrero 10, 1920, 6. The relation between eucalyptus and soil erosion in Bogotá seems to be marginal; however, in India and South Africa eucalyptus was used to tackle poor soil conditions, see Bennett, "The El Dorado of Forestry."
47 Portocarreño, "El gran problema de Bogotá," 6a.
48 "La arborización como medio de aumentar las aguas. El sistema parece ineficaz. Declaraciones de altas autoridades científicas," *El Tiempo (Bogotá)*, April 3, 1923, 6a.
49 "La arborización como medio de aumentar las aguas," 6a.
50 Kenneth Thomson, "The Australian Fever Tree in California: Eucalyptus and Malaria Prophylaxis," *Annals of the Association of American Geographers* 40 (June 1970): 237.
51 "La arborización como medio de aumentar las aguas," 6a.
52 Ibid.
53 Arnold Schultze, "Los perjuicios que causa el eucaliptus," *Revista de Industrias (Bogotá)*, August 1924, 59–60.
54 Rainer W. Bussmann et al., "Astonishing Diversity—the Medicinal Plant Markets of Bogotá, Colombia," *Journal of Ethnobiology and Ethnomedicine* 14 (2018), https://doi.org/10.1186/s13002-018-0241-8.
55 Darío Pérez and Laura Catalina Matiz-Guerra, "Uso de las plantas por comunidades campesinas en la ruralidad de Bogotá D.C., Colombia," *Caldasia* 39, no. 1 (June 27, 2017): 68, https://doi.org/10.15446/caldasia.v39n1.59932; Vargas, Mejía, and Guerrero, "Leña como combustible doméstico en zonas rurales de Usme, Bogotá." *Informador Técnico* 75 (December 2011): 30–39.
56 Sandra Liliana Rojas, "Structure and Floristic Composition of the Vegetation Undergoing Ecological Restoration in the Eastern Hills of Bogotá (Colombia)," *Caldasia* 39 (January 2017): 124–39.

A disciplined floristic island in the Tropical Andes

In the early republican Bogotá, the relationships between people and plants in the built environments occurred in private spaces beyond walls. While housemistresses from the elite transformed domestic patios into gardens planting carnations and lilies in broken pots, their servants, many of them women, grew, cared and manipulated edible and medicinal plants in the solares. Outside these private real enjoyed only by some members of the elite, people transiting across the city rarely have access to places decorated or cultivated with plants. The encounter with plants experienced by people without access to solares or patios unfolded in open spaces. There, spontaneous vegetation resulting from the botanical blend of native and early introduced species thrived. Some of them were edible plants and passers-by disputed birds and other free-wandering animals rich in sugar fruits of *uchuva, curuba* or blackberries growing in abandoned solares or unbuilt plots. In the meantime, this 1870s city of 50,000 inhabitants functioned thanks to a series of hidden relationships between people and plants unfolding beyond the built environment. Some people at the bottom of the social scale, who were usually of indigenous ancestry, travelled to the nearby forest and páramos, where they extracted firewood, lumber, fibbers and construction raw materials without which this pre-industrial city could not function. Hence, nineteenth-century Bogotá was a place where the relationships between people and plants were socially and spatially segregated. While women for the elite cultivated ornamental plants, people at the social margin had particular encounters with spontaneous vegetation in the city or established relationships with native species growing in places where housemistresses would never be by themselves.

The full independence from Spain in 1819 and the following integration of Colombia into the international capitalist trade radically changed the ways of imagining, inhabiting and building the city. This meant a transformation in the traditional and segregated botanical interactions with plants hitherto existing. The continuous introduction of technological innovations propelled by the growing world industrialisation brought the light

DOI: 10.4324/9781003208167-15

of modernity to the city. People from all around the country in formation, many of which had been dispossessed after several decades of civil wars, moved to the city appealed by this light. In a matter of decades, the city population doubled, reaching 100,000 by 1890. However, compared with other Latin American capitals such as Buenos Aires, Bogotá's growth and adoption of capitalist social arranges was slow. By the end of the nineteenth century, this high-altitude city still bears the social and economic burden of its relegated colonial past, stressed by the catastrophic aftermaths of the successive civil wars. As a result, the urban infrastructure did not match the growing population, and many of those moving to the city ended up living on the streets or sharing the living space in small places on the outskirts of the city. Overcrowded, the modern city that was leading behind its colonial pasts was the scenario of several epidemic outbreaks that, despite the seminal medicine progresses, kill hundreds every time. The poor hygienic conditions would trigger a large-scale transformation in the ways of conceiving and using plants in the city.

Preoccupied with the physical and, therefore, 'moral degeneration' of the Bogotanians and largely inspired by Georges-Eugène Haussmann's renovation of Paris, the emergent city's elites decided to endow the city with parks and public gardens. Guided by the bad airs medical mindset, open spaces with plants would be oxygenated places where people could go and enjoy fresh air free of miasmas. Additionally, in the mind of elites, gardens and parks were suitable places where the new labour class could enjoy hygienic strolls with family keeping men away from alcohol and the so-called *chicherias* [pubs]. Consequently, in 1880, plants that had been cultivated only in private spaces jumped outside, thus becoming a key raw material in the construction of the emerging urban public space. But, different to what had occurred in other cities such as Mexico City, where pre-Columbian horticultural practices had survived and adapted to the colonial and republican spaces of power representation (i.e. Chapultepec Garden), in Bogotá, the gardening and horticulture practices were of tiny dimensions, and in any case, always occurred in a domestic scale. It was impossible for a city without gardens to produce gardeners or people who specialised in the creation of public spaces; as a consequence, and despite the partial involvement of some foreigners, the creation of gardens in Bogotá was undertaken by self-taught gardeners. Casiano Salcedo, a gardening enthusiast, was appointed to green the city. Salcedo had learnt botany and gardening in books, and from books, he extracted the information and the landscape models he established in Bogotá. Because most of the gardening books were produced in France, he learnt French, and afterwards, he established contact with nurseries from where he introduced Parisian plants to match what he had seen in these books. Although Salcedo was more aware of the ornamental potential of Andean plants than housemistresses in cultivated the same kind of plant since the Spanish arrival, the beginning of the plant ornamentation in the city was mostly based on foreign species.

The introduction of plants to Bogotá linked to the creation of open public spaces made part of the large and usually underestimated nineteenth-century global reconfiguration of plant species linked to urban life. During the last decades of this century, the growth of the urban world made it necessary to enrich the city with plants that, while cleaning the air, provide citizens with a particular aesthetic scenery. Ornamental plants were at the core of the city's revegetalisation. Therefore, an entanglement of plant hunters, nurseries and botanical gardens played together to search, extract and place wild plants into an industrial-like process that turned them into ornaments to be sold and exported alongside the urbanistic models. Despite the isolation of Bogotá, the international market of plants soon reached this city. Plant hunters and landscape designers such as the French Édouard André and the British Robert Thomson, as well as plant merchants such as Mr McLane from Rochester's Live Plants Company, represent the two tips of an international business of enormous dimension that placed Latin American countries as a raw material supplier within the international trade. That is to say that if, on the one hand, plant hunters extracted orchids and anthuriums from wild Andean forest, at the same time, plant merchants sold ornamental species to the urban elites dreaming of having small Paris in the tropical mountains of South America.

The creation of public green space in Bogotá meant a reorganisation of the thus far rather stable urban flora formed by native plants coexisting with those brought with the Spanish conquest. However, starting in the twentieth century, the constant introduction of ornamental plants such as enset (*Ensete ventricosum*), araucarias and different kinds of flowering plants radically changed the botanical repertoire of the city. Because of the lack of economic sources or because of the constating ecological conditions of Bogotá that include large temperature oscillations, many of the recently imported plants just disappeared. Other species simply found suitable conditions in the city and soon naturalised, thus adding a new set of spontaneous plants to be recruited in every degraded area and open space in and around the city (i.e. *Impatiens walleriana*). Some naturalised, however, thrived too well and, after some decades, became invasives. This is the case of the common gorse (*Ulex europaeus*), which was mentioned by sources as a common nineteenth-century ornamental plant, would end up forming dense colonies, thus halting the development of today's Andean ecosystems around the city.[1] The input of botanical material experienced in Bogotá in the transition to the twentieth century matched with the physical transformation of rural-like places such as riverbanks and solares, which had hosted various ruderal and edible species. Therefore, a city with more gardens produced with introduced species that at the same time urbanised residual spaces whereby spontaneous plants had historically thrived soon transformed Bogotá into a floristic island different in species than its surroundings.

The spatial definition of Bogotá as a floristic island was stressed after the introduction of the forestry industry. Despite the enormous diversity of the ecosystems around the city, structural ecological conditions linked to historical overexploitation prevented nearby forests from covering the demands of wood imposed by a growing population. Until late in the century, and following a colonial tradition, lumber and firewood were introduced into the city from neighbour lowland valleys several kilometres away. However, this was cumbersome and costly. As a way to tackle the growing need for lumber and firewood linked to the growing demands of the capitalist city, the self-taught gardener Casiano Salcedo planted the first *Eucalyptus globulus* seed in Bogotá in the decade of 1880. Although, in the beginning, this Australian tree was almost exclusively used in the ornamentation of parks and avenues, by the first decades of the twentieth century, its potential as raw material in the creation of artificial forests was fully unfolded. Thus, when in 1918, massive eucalyptus monocultures were implemented in the eastern mountains, and the city would finally have a nearby source of wood. Nevertheless, this plantation meant that the ecological succession of these mountains was radically relented. Native species simply could not compete with the growth rate of the recently introduced tree and were confined to steeped areas not reached by the plantations. Additionally, this 'artificial forest' limited the access of people to the mountains. Without access to the resources found in nearby ecosystems, straw and fibber collectors, for example, had to travel even farther or abandon their source of income. Consequently, the establishment of the eucalyptus forest around Bogotá meant the definition of the city as a floristical island where emergent interaction with plants developed.

This floristical island which was produced after the creation of modern green spaces, the introduction of plants and the degradation of nearby environments, promoted ways of relating to plants completely different to what has existed in the history of the city. Whereas the relationships with plants in places such as solares, patios, or nearby environments were products of the city's social and cultural heritage from the Muisca society and Spanish colonisers, the modern green spaces propelled an inverse construction. This means that rather than being created as a natural expression of historical relationships with plants, parks and public gardens were built first, and it was not until later that the mediators in the guise of doctors and gardeners induced and promoted new human–plant interactions. The result of this intermediation was a significant estrangement between people and plants in the city, which in turn stressed the lack of gardening tradition in the city.

The estrangement between people and plants confirms how modernisation was a huge process of delegation of many aspects of daily life. For example, those who had previously worked as farmers became labourers who delegated the production of their food to others, as was the case of Mexican *tortillas* and Colombian *arepas*.[2] The same happened with the production and repair of their clothes, the construction of their houses and the education of their children. Similarly, the previously intimate relationships between humans and plants became based on the already mentioned local characters, which in turn became mediators between people and plants in Bogotá. For instance, the gardener emerged when the spontaneous bushes and weeds of the residual and pre-modern areas were replaced by organised tree-lined avenues. Utterly unnecessary in the rural world, the figure of the gardener was the epitome of urban life. Gardeners, then, emerged as mediators, first as self-taught figures such as Casiano Salcedo and later as an organised guild headed by professional gardeners such as José Fabi. The same occurred with the non-timber forest products (NTFPs) obtained in the nearby mountains, some of which, such as the spoons made from *totumos*, were replaced by imported metal utensils as a manifestation of industrial innovations. In the same vein, people embodying the fictional character of *Bruna the Charcoal-Woman* were replaced by others carrying technical knowledge, who transformed the mountains of Monserrate into a monoculture of eucalyptus. Consequently, the whole phenomenon of delegation implied in the creation of the urban flora of Bogotá meant a transition from ethnobotanical to technobotanical practices.

As public spaces, gardens and parks entailed a multidimensional social dynamic that extends beyond the people responsible for their design and maintenance, the delegation in the relationships between people and plans extended beyond the figure of gardeners. Hence, new green spaces gave life to garden guards locally known as *celadores*. Although usually relegated by urban historiography, the figure of the celador was as important as the gardens in the construction of the urban flora. Whilst the latter administrated and organised the plants in the space, the former regulated the conflict and behaviours of people in relation to plants. This conflict emerged because the adoption and acclimation of ideas, plants and spaces in Bogotá was not a strict process of replacement. Accordingly, the modernisation of the city's green spaces was rather a gradual superimposition whereby modern and colonial ideas around the role of plants co-existed for some time. These encounters between contrasting ways of living with plants brought about an inevitable conflict.

In the eyes of the elites, parks and public gardens were places for hygiene strolls and social divertimento; however, when they met with the others there, they soon discovered that green spaces meant other things to the rest of the people. The benches were free-of-rats beds for the ill and vagabonds;

the lawns were playgrounds for children, trees were organic urinals, and their trucks were a source of firewood for the poorest. These differences in the ways of understanding the social function of green spaces turned them into places of contestation and were, in not a few instances, tempted the ruling class to forbid access to these spaces for the poor or ills. However, the attempts to ban the entrance of marginals to parks and public gardens never materialised. It was simply too risky for the elites to expel the labour class from parks. They were afraid that in doing so, they would end up in pubs, thus omitting their responsibilities at home and in the workplace. Celadores then were responsible for instigating a series of social behaviours in their visitors. For example, they were in charge of suffocating the appearance of pre-modern practices in modern parks when people lacking firewood were chopped ornamental trees. The presence of celadores and a series of norms aimed at controlling the relationships with plants revealed how, by the end of the nineteenth century, the thus far spontaneous relations with plants were replaced by regulated and disciplined forms of understanding, use and performance with the newly created botanical realm.

In many instances, the power relationships around the relationship with plants were subtle. This is evident when as an alternative to the repressive and punitive activity of celadores, the city's elite decided to focus on the popular education structured around the symbolic figure of the tree. Didactic materials exalting the figure of the tree were published and used in schools as part of the pedagogic toolbox. Anthems, prayers and poems addressing the importance of trees were usually read in the classroom. This green proselytism reached its epitome in 1912 during the tree's day. Celebrated the same days of the national independence (the 20th of July), this tree celebration aimed to cultivate the love for trees and for the nation within the city's children while, at the same time, they planted thousands of trees with their small hands in the city's green spaces. Nonetheless, the symbolic use of the tree was not a new instrument in the creation of disciplined citizens. The grandparents of the kids attending the trees' day had experienced something similar. In 1813, the newborn republic had chosen a tree to symbolise the independence of Spain and used it as a patibulum whereby a 13-year-old slave would be executed to represent in front of the masses the power of the new republican order. However, the independence from Spain, and the later integration of Colombia into the international trade, did not radically change the conditions of life for people in Bogotá. Therefore, sons and daughters of poor mestizos and indigenous did not make it better than their ancestors. Even more, the social conditions of Bogotá worsened with the increase of the population. More people in an unequal society are soon prone to the emergence of crime, orphanages and diseases. Consequently, orphanages, jails and hospitals were established in the city, and there also, plants played a key role when used as a therapeutic element, and ornamentation played a special role in building healthy, strong and disciplined citizens. As demonstrated with the tree's day, the liberty tree and the therapeutic use of plants in the rehabilitation of marginal people, plants but especially trees, played a key role in the construction of a collective identity formed by disciplined and healthy people.

This book has shown how the creation of modern gardens and parks in a highly biodiverse city such as Bogotá entailed the creation of a modern flora characterised by being richer in plant species than their city surroundings and by entailed new disciplined interactions with plants. In doing so, this book contributes to expanding the urban and gardening history of tropical Latin America. However, there is still too much to understand about the negotiations, social tension and ecological consequences involved in the creation of tropical urban natures. These investigations can be taken from at least three perspectives. First, future research should inquire about the role of plants as part of modern biopolitics. The use of plants as non-human actors involved in the creation of social discipline in the case of the San Jose orphanage's gardens is just an instance of the roles of plants in the modern biopolitical structure. Given current academic ideas in the more-than-human geographies on the social roles of non-human actors, research on the role of plants as organic devices of discipline can offer a different perspective of the interconnected nature of humans and non-humans, which are usually addressed from a conflict-blind perspective.[3]

Second, this book has briefly addressed the international trade that engendered a completely new flora in the nineteenth-century Bogotá. However, this is just a very initial attempt to understand how modernisation made possible an exchange of floras in different regions of Latin America. In contrast to the extensive investigations describing and explaining the transatlantic Columbian exchange, almost nothing has been done thus far to reveal the exchange of flora triggered by the consolidation of urban life in Latin America and the expansion of the capitalist system in the region. For example, the commercial networks between Europe and Latin America researched herein require a more in-depth analysis. In contrast to the rather well-studied exploitation and looting of orchids in Latin America by individuals devoted to supplying the growing European demand for rare flowers, the introduction of plants into nineteenth-century Latin American cities is still a theme that needs further development. Research addressing the introduction of plants as a consequence of urban growth in Latin America can inform us not only about the trade of exotic plants in Latin America but also how significant changes in the natural distribution of creating species were triggered by ideas about plants that, for example, preferred temperate species over native plants growing in highly biodiverse countries.

Finally, despite the methodological limitations posed by the lack of adequate archives, studies that examine the historical transformation in the use of plants in urban environments must also consider cities that encompass the social and ecological diversity present in Latin America. Academia has yet to fully understand the environmental processes that have occurred in mountain cities, desert cities, savanna cities or forest cities. For instance, regarding the latter, although these areas were usually in a peripherical position in relation to the centres of power such as Bogotá, investigating cities such as Leticia in the Colombian Amazon and Quibdó in the Chocó biogeographical region poses significant methodological difficulties; however, such studies could help to enrich the sources that inform the modernisation of the human–plant interactions by using alternative approaches based on oral traditions and other traditional knowledge. Such studies in peripherical cities with completely different socioecological realities could expand our poor understanding of other ways of building the city that is not necessarily influenced by modern, western(ised) guidelines.

Notes

1 The common gorse represents an ecological issue in today's Bogotá. Although some scholars have observed that this plant was introduced in the 1950s, there is evidence of its use as ornamental as early as 1890, see Gómez-Ruiz, Lindig-Cisneros, and Vargas-Ríos, "Facilitation among Plants: A Strategy for the Ecological Restoration of the High-Andean Forest (Bogotá, D.C.—Colombia)," 267–75.

2 Lise Sedrez and Regina Horta Duarte, "The Ivy and the Wall: Environmental Narratives from an Urban". Claudia Leal, John Soluri, and José Augusto Pádua (eds.), *A Living Past: Environmental Histories of Modern Latin America* (New York: Berhahn Books, 2019) 138–162.

3 The intersection between biopolitics and plants has recently begun receiving scholarly attention; however, research efforts have been almost entirely directed toward biopolitical ideas as an explanation of the historical underestimation of plants against animals. Matthew Hall, *Plants as Persons: A Philosophical Botany, SUNY Series on Religion and the Environment* (Albany: State University of New York Press, 2011); Jeffrey T. Nealon, *Plant Theory: Biopower and Vegetable Life* (Palo Alto, US: Stanford University Press, 2015); This book offers a different interpretation of the Foucault biopolitics by using typical examples such as hospitals or prisons to reveal the critical role of plants as biopolitical devices in the modern and urban control of human beings, see also Michel Foucault, *Discipline and Punish: The Birth of the Prison* (London: Allen Lane, 1977).

Index